O
IN
INTRODUCTIONS

# INDIAN NUCLEAR POLICY

The Oxford India Short
Introductions are concise,
stimulating, and accessible guides
to different aspects of India.
Combining authoritative analysis,
new ideas, and diverse perspectives,
they discuss subjects which are
topical yet enduring, as also
emerging areas of study and debate.

For more information, visit our website:
https://india.oup.com/content/series/o/
oxford-india-short-introductions/

OXFORD
INDIA SHORT
INTRODUCTIONS

# INDIAN NUCLEAR POLICY

HARSH V. PANT

YOGESH JOSHI

OXFORD
UNIVERSITY PRESS

**OXFORD**
UNIVERSITY PRESS

Oxford University Press is a department of the University of Oxford.
It furthers the University's objective of excellence in research, scholarship,
and education by publishing worldwide. Oxford is a registered trademark of
Oxford University Press in the UK and in certain other countries.

Published in India by
Oxford University Press
2/11 Ground Floor, Ansari Road, Daryaganj, New Delhi 110 002, India

ISBN-13 (print edition): 978-0-19-948902-2
ISBN-10 (print edition): 0-19-948902-5

ISBN-13 (eBook): 978-0-19-909383-0
ISBN-10 (eBook): 0-19-909383-0

Typeset in 11/14.3 Bembo Std
by The Graphics Solution, New Delhi 110 092
Printed in India by Replika Press Pvt. Ltd

# Contents

# Preface

India has travelled a long distance from being a nuclear pariah to a de facto member of the nuclear club. As India continues to search for its complete integration into the global nuclear order, this book explores the trajectory of Indian nuclear policy from the early days since Independence to the present. In so doing, it hopes to underline key debates, both policy and conceptual, that have shaped this trajectory.

This is a huge subject but the constraints of Oxford India Short Introductions (OISI) series forced us to be more focused and disciplined in our approach. We would like to thank the editors at Oxford University Press for patiently working with us on the project till completion. Thanks to Observer Research Foundation, New Delhi, where both of us were based during the conception and completion of the project. We are especially thankful to Gaurav Sharma who generously shared his PhD dissertation on India and nuclear disarmament with us, as well as to Akshay Ranade and Ketan Mehta for their help in preparing the bibliography. This book would not have been possible without the time so many members of the Indian scientific, defence, and foreign policy establishment gave us, sharing their views and ideas. Finally, we owe a deep debt of gratitude to our respective families whose support has kept us afloat in more ways than one.

This book is dedicated to our teachers who have shaped our intellectual trajectories. Nothing would have been possible without them.

# Abbreviations

| | |
|---|---|
| AEC | Atomic Energy Commission |
| BARC | Bhabha Atomic Research Centre |
| BJP | Bharatiya Janata Party |
| CBM | Confidence-Building Measure |
| CIRUS | Canada India Reactor Utility Service |
| CMD | Credible Minimum Deterrence |
| CTBT | Comprehensive Test Ban Treaty |
| DAE | Department of Atomic Energy |
| DRDO | Defence Research and Development Organisation |
| ENDC | Eighteen Nation Disarmament Committee |
| FMCT | Fissile Material Cut-off Treaty |
| GE | General Electric |
| IADA | International Atomic Development Authority |

| IAEA | International Atomic Energy Agency |
| IAF | Indian Air Force |
| IGMDP | Integrated Guided Missile Development Programme |
| IMF | International Monetary Fund |
| JIC | Joint Intelligence Committee |
| kg | Kilogram |
| km | Kilometre |
| MEA | Ministry of External Affairs |
| MIRV | Multiple Independently Targetable Re-entry Vehicle |
| MoD | Ministry of Defence |
| MoF | Ministry of Finance |
| MTCR | Missile Technology Control Regime |
| MWT | Megawatt Thermal |
| NAI | National Archives of India |
| NCA | Nuclear Command Authority |
| NDA | National Democratic Alliance |
| NFU | No First Use |
| NMML | Nehru Memorial Museum Library |
| NNPA | Nuclear Non-proliferation Act |
| NNWS | Non-nuclear Weapon States |
| NPT | Non-Proliferation Treaty |
| NSAB | National Security Advisory Board |
| NSG | Nuclear Suppliers Group |
| PMO | Prime Minister's Office |
| PNE | Peaceful Nuclear Explosion |

| R&D | Research and Development |
| RAPS | Rajasthan Atomic Power Station |
| RAW | Research and Analysis Wing |
| SFC | Strategic Forces Command |
| SLBMs | Sea-Launched Ballistic Missiles |
| TAPS | Tarapur Atomic Power Stations |
| TIFR | Tata Institute of Fundamental Research |
| TNWs | Tactical Nuclear Weapons |
| UK | United Kingdom |
| UKAEC | United Kingdom Atomic Energy Authority |
| UN | United Nations |
| UNDC | United Nations Disarmament Commission |
| UNGA | United Nations General Assembly |
| US | United States |
| USAEC | US Atomic Energy Commission |
| USSR | Union of Soviet Socialist Republics |

# Introduction

India's first prime minister, Jawaharlal Nehru, laid the foundations of an elaborate atomic energy programme in April 1948, less than a year after India's independence. Yet, it took Indian decision-makers more than 50 years to declare the country a nuclear weapon state in May 1998. For the first five decades of India's independence, its nuclear policy remained highly ambivalent. On the one hand, it continuously strived to develop all aspects of atomic energy, including an explosive nuclear capability. On the other, it maintained a moral and political commitment to peaceful uses of nuclear energy and vouched vehemently for nuclear disarmament. This ambivalence was partly situated in the duality of the atom: it could be used for peaceful as well as destructive purposes. It was also a result of the ideological dimension of Indian foreign policy and India's relative lack of material power. In fact, this ambivalence continued despite changes in India's security environment and transitions in domestic politics and individual leadership. Yet, over the years, the weight of all these factors shifted continuously towards a point where ambiguity could not be sustained further. Therefore, rather than consciously choosing a nuclear weapons path, India was almost forced into being a nuclear weapons state. And once India crossed the nuclear rubicon in May 1998, it managed to transform itself into a major nuclear power. Also, within two decades of this event, India is not only making rapid technological advancement in its nuclear capability but has also managed to get accommodated in the global nuclear order. It is the only nuclear weapon state apart from the five established nuclear powers which is legally allowed to have both a nuclear weapons programme and a civilian nuclear energy programme. India is determined towards full integration in the global nuclear order, evident in its bid to become a member of technology control regimes such as the Nuclear Suppliers Group (NSG). This transition in India's nuclear identity has been accompanied by its transformation into a major economic power and underlines a pragmatic turn in its foreign policy thinking since the end of the Cold War.

This book argues that India's engagement with the atom is unique in international nuclear history and politics. This uniqueness emanates from three distinct features of India's nuclear policy. First of all, India's nuclear weapons programme has been an offshoot of its atomic energy programme. All other nuclear weapon states first pursued nuclear weapons before they turned their gaze to peaceful uses of atomic energy. This unique path also makes India the only country to have debated the decision to pursue or not to pursue the nuclear weapons path extensively, nearly for five decades: 'no nation has debated more democratically than India whether to acquire or give-up nuclear weapons' (Perkovich 1999: 447). Second, India's nuclear history disproves the linear model of nuclear weapons proliferation where insecurity vis-à-vis a bigger and hostile nuclear power is the principal source of a state's motivation to pursue nuclear weapons, as was the case with the Soviet Union, China, and, to a certain extent, both the United Kingdom (UK) and France. The Indian case, interpreted correctly, disproves the linear model of nuclear proliferation as it was Pakistan, rather than China, that was the most important reason for India to go nuclear (Joshi 2017a). It took India 10 years to respond to the Chinese nuclear test in October 1964; and when India first tested a nuclear device in 1974, it called it a peaceful nuclear explosion (PNE) and followed a policy of 'nuclear refrain'. It deliberately eschewed converting its nuclear explosive capability into an active nuclear weapons programme for at least another decade, till the time Pakistani nuclear weapons programme matured into a full-blown threat to the Indian state by the mid-1980s. Finally, more than any other major nuclear weapons power, India has exhibited a strong moral and political revulsion to nuclear weapons. This aversion to nuclear weapons emanated out of the ideological underpinnings of India's freedom struggle and inclinations of its post-independence political leadership. Even when it mastered all aspects of atomic energy, including an explosive capability, the quest for disarmament remained potent. A strong belief existed among Indian decision-makers that a nuclear-free world would be more advantageous to India's security than a world teeming with nuclear weapon states. This reason also explains India's continuous commitment to non-proliferation even when, in the past, it became one of the biggest targets of the international non-proliferation regime. Even today, India remains committed to the idea that nuclear weapons are political tools rather than instruments of warfighting. Yet, its decision to ultimately become a nuclear weapon state attests to the fact that it cannot remain 'immune' from the application of power on the pretext that the power principle in international politics is 'uncivilised' (Krishna 1984: 286).

4

To explain India's nuclear policy and practice, this book follows a linear historical narrative beginning in 1947. It has been divided into five major chapters. Chapter 1 underlines the early phase of India's engagement with atomic energy. It delves into how India's nuclear pioneers—Prime Minister Nehru along with Homi Bhabha—laid the foundations of atomic energy research, created important scientific institutions, and formulated India's response to global nuclear debates of the time. Chapter 2 focuses on India's nuclear policy in the aftermath of the Chinese nuclear test in October 1964 till India's PNE in May 1974. Chapter 3 delineates India's nuclear behaviour in the post-PNE era. Between 1974 and 1984, India's nuclear policy faced great complexity, especially that of the emerging non-proliferation regime and the threat of a nuclear Pakistan. Chapter 4 traces the final years of India's journey into becoming a full-fledged nuclear weapon state. Between 1984 and 1998, as the threat from Pakistani nuclear weapons became more and more apparent and the non-proliferation regime strengthened, India's nuclear ambivalence became unsustainable. The then prime minister, Atal Bihari Vajpayee, ended this five-decade-old policy with the May 1998 nuclear weapon tests. The last chapter, Chapter 5, traces India's transformation into a major nuclear power in the last two decades. It explains major

5

continuities and changes accompanying India's nuclear policy in the aftermath of the 1998 tests.

This historical narrative, however, is situated in a theoretical framework which focuses on four major factors as guiding India's nuclear policy and practice: status, security, domestic politics, and the role of individuals. The book underlines how all these factors, sometimes solely and mostly in interaction with each other, shaped India's nuclear trajectory, its policy choices, and also their consequences, both negative and positive. It uses not only the existing secondary literature available on India's nuclear policy but also makes extensive use of primary documentation, including the recently declassified archival documents, to illustrate some important aspects of India's nuclear policy.

## Debating the Causal Variables

The aforementioned four factors have played a key role in shaping India's nuclear policies. India's quest for major power status is a dominating theme in India's nuclear journey. The advent of nuclear weapons transformed great power politics and bestowed upon its masters the ultimate military capability. In the post-Hiroshima period, nuclear weapons increasingly became the benchmark for states to attain great power status. As the Cold War unfolded, this was further formalized with all nuclear weapon states being the permanent members of the United Nations Security Council. For Indian decision-makers, starting with Prime Minister Nehru, making India a major global power was a fundamental foreign policy goal. India's dogged pursuance of atomic energy was, therefore, one of the means to attain this foreign policy objective. At least, the attempt was to make India capable of mastering the atom both peacefully and to reach the threshold where it can also unleash its destructive potential. Yet, India's national identity as a major power also exhibited certain exceptionalism. This exceptionalism was built upon an idea of 'moral superiority' stemming from India's non-violent freedom struggle, differentiating it from other major powers whose primary claim to status lay in their military might. Such exceptionalism allowed India, especially under Prime Minister Nehru, to be the moral policeman of the world, even when materially it remained confined to the backwaters of international politics. India's national identity as a major power was therefore a combination of classical great power politics on the one hand, and its moral exceptionalism on the other. These two notions of status, however, pulled India's nuclear policy in two opposite directions: one pushing it towards a nuclear weapons capability and the other forcing it to be the leading voice on nuclear disarmament.

If status was a driving factor especially during the early years of India's nuclear journey, its security requirements provided her the most potent motivation to pursue a nuclear weapons programme. Nehru's interest in developing atomic energy research provided India a hedge against future uncertainties where it may have to face a nuclear-armed adversary. When China conducted its first nuclear test in 1964, India was one of the most advanced nuclear technology states in the Third World. In the wake of the Chinese tests, India began developing a nuclear explosive capability under the guise of PNEs. This process of developing the 'nuclear option' was a result of insecurity engendered by the Chinese tests and the domestic pressure upon the Indian decision-makers to counter it. Even while developing the 'nuclear option', Indian decision-makers pursued diplomatic strategies to contain the fallout of the Chinese bomb. Between 1964 and 1974, Indian security improved due to indigenous development of its conventional military capability, the Sino-Soviet rift, and the domestic political chaos in Mao's China. In fact, when India conducted its peaceful nuclear explosion in 1974, the nuclear threat from China was considered by Indian decision-makers to be negligible. China, however, was a long-term threat and therefore remained a factor in India's ambivalent attitude towards nuclear weapons. By the late 1970s, emergence of a

Pakistani nuclear programme became the major concern of Indian decision-makers. Unlike China whose nuclear decisions were motivated by Cold War politics and the quest to deter the Soviet Union and the United States (US), Pakistani nuclear programme had no other motive but to blackmail India and to pursue its revisionist agenda in Kashmir and beyond. The fact that it was technologically supported by Beijing also created an impression of a nuclear siege by two hostile adversaries. India's reaction to the Pakistani nuclear programme was, therefore, drastically different than was the case with the Chinese nuclear programme. In the late 1980s, as Pakistan acquired a nuclear weapon capability, India raced towards operationalization of its own nuclear deterrent. By 1998, when India finally declared itself to be a nuclear weapon state, security had become the most dominant factor in India's nuclear calculus.

The other prominent factor shaping India's nuclear policy has been at the level of domestic politics. Being a pluralistic democracy, nuclear policy has been not only vigorously debated in the Indian Parliament but also engendered different schools of thought on India's nuclear choices, ranging from complete renunciation of the nuclear weapons to building the bomb, from the very beginning. These schools of thought have dominated India's nuclear policy discourse depending upon its external circumstances and the internal power structures. For example, the right-wing political parties such as the Jan Sangha and the Swatantra Party, and its progenies such as the Bharatiya Janata Party (BJP), have always been pro-bomb. The fact that these pro-nuclear parties remained on the margins of political power during the first 50 years of India's independence is critical to understanding India's nuclear ambivalence. Similarly, the Indian National Congress, which dominated the political scene during the first five decades, had some inherent resistance in openly embracing nuclear weapons given its ideological history, and also the momentum of the policies enunciated under Prime Minister Nehru where disarmament had become a prominent foreign policy goal. Yet, there was a political consensus to maintain India's nuclear option and to steadfastly defend its strategic autonomy. India's resistance to the diktats of the non-proliferation regime engendered out of this political consensus. India's pushback against nuclear non-proliferation treaties, such as the *Non-Proliferation Treaty* (NPT) and the Comprehensive Test Ban Treaty (CTBT), therefore enjoyed popular domestic support. Another factor in domestic politics is the role (or the lack of it) certain institutions have played in steering India's nuclear policy. The nuclear scientific enclave has played an active role as the expertise of civilian nuclear scientists provided them considerable influence in charting India's nuclear choices. The military, on the other hand, has remained at the margins of India's nuclear decision-making.

Finally, the role of key individuals in charting India's nuclear trajectory has been equally consequential. During the first two decades, Prime Minister Nehru and Homi Bhabha singlehandedly decided India's nuclear trajectory. They laid down the foundations of its atomic energy programme, created institutions, and pronounced the thrust of its nuclear policy. This also led to concentration of nuclear decision-making in two institutions: the Prime Minister's Office (PMO) and the Department of Atomic Energy (DAE). The prime minister and the secretary of the DAE (to a lesser degree), therefore, remained the two most important individuals in deciding India's nuclear policy. Depending upon the individual convictions of the prime minister and the secretary, DAE, India's nuclear policy shifted gears. The stark contrast between the thinking of Homi Bhabha and Vikram Sarabhai was a case in point, and so were the ideological convictions of leaders such as Lal Bahadur Shastri, Morarji Desai, and Atal Bihari Vajpayee. As we will see in subsequent chapters, the prominence of individuals in India's nuclear policy often lends it an idiosyncratic character.

The sources of India's nuclear policy can be traced to an interaction among the previously mentioned four variables, with one sometimes dominating the other depending upon India's external environment, the constellation of domestic power structures, and the ideological and personal convictions of individual leadership. The consequences of India's choices have also been equally pronounced.

## India's Nuclear Choices and Their Consequences

India's nuclear policy before 1998 can be divided into three major themes: nuclear ambiguity, selective defiance of the global nuclear order, and a crusade against nuclear weapons. Post–1998, however, India has followed a policy of nuclear certainty, where it openly accepts the logic and necessity of nuclear weapons, and in the process has become a 'normal' nuclear power. Yet, it maintains both a restrained and responsible nuclear profile. However, both these phases of India's nuclear policies have had certain consequences not only for its nuclear programme but also for its national security and foreign policy.

The policy of nuclear ambiguity where India maintained a peaceful exterior while developing threshold nuclear capability neither helped its security requirements nor did it fulfil its quest for major power status. Nuclear reluctance, in fact, costed her a seat in the global nuclear order which was first enshrined in the NPT of 1968. If India would have gone nuclear before the NPT was finalized, it would have been considered as a nuclear weapons power and would have cemented its major power status in the global nuclear order (Trivedi 1975). This nuclear reluctance was in part inspired by India's resource crunch; the era in which India suffered its worst food crisis did not provide its decision-makers the luxury to pursue a full-fledged nuclear weapons programme.[1] However, if the Indian state would have decided to go nuclear, resource-crunch alone could not have halted its nuclear steps. It is therefore important to underline that India's quest of 'moral superiority' and its inclination to prove that it was an exceptional major power was an equally important factor behind its nuclear reluctance. Yet, in the realm of international politics where military power remained the *ultima ratio* in inter-state interaction, such nuclear morality found no takers. In hindsight, India's moral leadership was only an outcome of Nehru's exceptional political leadership and his recognition as a statesman of international repute. With his demise,

---

[1]  For a detailed discussion on India's policy choices after the Chinese nuclear test, see Joshi (2017b).

India neither had a moral nor a material claim to major power status in international politics. Its nuclear reluctance turned out to be an additional burden.

However, India could not renounce its strategic autonomy and give up its nuclear option. It, therefore, defied the global nuclear order, which it considered both unsustainable and discriminatory. It first refused to sign the NPT in 1968, and then conducted a PNE in May 1974. Afterwards, it became the principal target of the nuclear non-proliferation regime, resulting in technology denials which crippled its nuclear energy programme. If its nuclear reluctance stymied its quest for a major power status, its defiance of the non-proliferation regime only brought her additional pain in the form of economic and technological sanctions. This defiance was selective insofar India's nuclear policies confirmed to the larger principle of non-proliferation. Even when it publicly supported the need for technological cooperation in peaceful uses of nuclear energy, in private it refused to share its nuclear technology and material with other states lest it would lead to proliferation of nuclear weapons. However, this responsible nuclear behaviour did not bring her any respite. On the other hand, open proliferators like China were not only accommodated in the global nuclear order but were allowed to build up their nuclear arsenal.

From Nehru's call for cessation of nuclear tests in 1954 to Rajiv Gandhi's appeal at the United Nations (UN) for complete elimination of nuclear weapons in 1988, disarmament remained one of the pivots of India's nuclear policy. Yet, its moral crusade against nuclear weapons was largely ignored by the great powers largely because India remained an ambiguous nuclear power. Without material power, its moral and ideological claim against nuclear weapons had no major appeal for great powers. India's diplomatic disappointment in creating a nuclear weapon-free world was a direct result of its nuclear reluctance. Until the tests of 1998, it remained on the margins of the global nuclear order.

The nuclear tests of 1998 ended not only India's nuclear ambiguity but also signalled its intention to deal with global politics in the language of power. Unlike years of nuclear ambiguity, it now embraced a policy of 'nuclear certainty' where nuclear weapons became fundamental to its security and identity, especially in a world where great powers were reluctant to embrace a nuclear weapon-free world. In hindsight, the consequences of a policy of nuclear certainty have been extraordinarily generous to India. Its rise as a major economic, political, and military power has taken place in the post-1998 period, even when many had argued that the tests would jeopardize its

15

economic transformation and bring her nothing more than international opprobrium. In fact, within a decade, India was accommodated in the global nuclear order. The Indo–US nuclear deal created an exception in the global nuclear order where India is the only non–NPT state to have a recognized nuclear weapon programme. The transformation in India's nuclear identity has been nothing less than phenomenal. In the last two decades, India has also marched rapidly in advancing its nuclear arsenal. Yet, it still remains a restrained nuclear power which considers nuclear weapons, primarily, as political tools rather than instruments of warfighting. If nuclear restraint is one aspect of its post–1998 nuclear policy, nuclear responsibility is another. On the one hand, it openly supports the principle of non–proliferation and actively contributes to efforts aimed at stopping further spread of nuclear weapons. On the other, this responsible nature has also been evident in its restrained response to the grave provocations by Pakistan since 1998. In crisis after crisis, where India faced either armed conflict or terrorism from Pakistan, it has acted with great restraint in dealing with Pakistani adventures.

Yet, in the coming times, Indian nuclear policy is likely to face complex challenges. First of all, India's accommodation in the global nuclear order will continue to be the focus of its nuclear policy. Even though the Indo–US nuclear deal granted India an exceptional status of being the only non–NPT nuclear weapon state with dedicated military and civilian nuclear programmes, this accommodation remains partial. India intends to become a full member of the global nuclear order, which is evident in its continuing quest for the membership of the NSG. However, India's efforts have not yet yielded the desired results. Future negotiations on India's nuclear status would require not only keen diplomatic acumen but also marshalling of India's economic, military, and technological influence. This process would also entail some give and take on the part of Indian leadership, especially on arms control treaties such as the CTBT and Fissile Material Cut-off Treaty (FMCT). India's emergence as a global nuclear power would also create frictions with the other Asian nuclear giant, China. As India's nuclear status provides her the platform to challenge China's hegemony in Asia and beyond, Beijing would remain cautious in accepting India as a legitimate member of the global nuclear order. It would continue to resist India's rise, as evident in China's opposition to New Delhi's NSG membership. Moreover, the tense equation between India and China over the latter's claims on Indian territory and her attempts to limit the former's influence in South Asia will make sure that the nuclear rivalry would remain potent. India is still some distance away from mastering a credible second-strike capability against China and therefore, the quest for a stable deterrent equation with Beijing would drive the technological advancement of India's nuclear forces. However, Pakistan would remain the prickliest of all challenges. Its penchant for nuclear risk-taking and her inclination towards first use of nuclear weapons has grave consequences for crisis stability in South Asia. How India ensures its security while facing a revisionist nuclear state prone to inciting crisis is a question which the Indian decision-makers will continue to grapple with. Lastly, India's nuclear identity would also be influenced by the political churning within. Historically, centre–right political parties such as the BJP have championed India's overt nuclearization and have found little traction with the moralistic tendencies of its Nehruvian past. With the coming of the Narendra Modi government, there has been an expectation that India's nuclear profile will undergo some important changes. The amount of diplomatic capital Modi has invested in India's NSG membership vis-à-vis the previous Congress-led government is a case in point. Domestic political changes are likely to be an important factor in shaping the future of India's nuclear trajectory.

Indian nuclear policy remains a work in progress. The way it evolves over the next few years will have an important bearing on India's global profile, as well as on the larger global non-proliferation regime.

# 1

# The Promise of the Atomic Age

India's tryst with the atom followed her tryst with freedom. Soon after India gained independence in August 1947, Prime Minister Jawaharlal Nehru initiated an atomic energy programme. If freedom from colonial rule provided India an independent identity, science and technology, in Nehru's worldview, provided an avenue for a rapid march towards modernity. Atomic research, at the time, was at the vanguard of this scientific modernity. Nehru was highly impressed by the prestige associated with this field of scientific enquiry. On the one hand, it would make India a leading scientific power among the Third World countries and on the other hand, it would also generate respect among the technologically advanced countries of the West. Indian political leadership had criticized the use of atomic weapons at

Hiroshima and Nagasaki and expressed a deep moral indignation for such weapons of mass destruction. Yet, Nehru also could not ignore the reality of a nuclear world: nuclear weapons provided a nation state the ultimate military capability. This ambiguity in Nehru's approach to the atom was to be a defining feature of India's nuclear policy for decades to come.

This chapter traces the beginning of India's nuclear polices, its atomic energy programme, and the nature of its nuclear intentions. The first section of this chapter throws light on the ideas of its nuclear pioneers: Prime Minister Jawaharlal Nehru and Homi Bhabha. These two individuals remained at the helm of India's nuclear policy till the mid-1960s and shaped not only the origins of India's atomic programme but also its development in the first two decades after India's independence. It also elaborates on how these nuclear pioneers initially conceptualized India's atomic programme, created institutions, and put in place mechanisms to harness the promise of the atomic age. The second section delineates the international dimension of India's nuclear programme and the cooperation agreements India signed with major nuclear powers such as the UK, Canada, and the US. India's role in shaping the debate on international control of atomic energy is the focus of the third section. It specifically deals with India's approach to the International Atomic Energy Agency (IAEA) and the system of safeguards on nuclear technology. The last section discusses one of the major pillars of India's nuclear policy during the first two decades of its existence: its approach to nuclear disarmament.

## India's Nuclear Pioneers and the Atomic Establishment

Nehru's interest in atomic energy was shaped by three ideological vectors. First, he was a man of 'scientific temper' (Nehru 1986: 409). He considered modernity as synonymous with advancement of science and technology. It was on the role of modern science and technology in India's future as a nation state that he sharply disagreed with Mahatma Gandhi, his political mentor. Gandhi saw in modern science the real origins of colonialism and in his book, *Hind Swaraj*, had severely critiqued the idea of 'modern civilization' primarily built upon Western advancements in science and technology. Nehru, on the other hand, claimed that it was the lack of scientific progress that allowed India to be colonized. His strong belief in the virtues of modern science led India's first prime minister to invest heavily in creating an environment where science and technology could prosper (Andersen 2010).

Second, if the promise of atomic energy animated Nehru, its destructive potential imbued in him a moral

21

aversion to nuclear weapons. As he argued in August 1947 during the Constitutional Committee debate, 'in essence today there is a conflict in the world between two things, the atom bomb and what it represents and the spirit of humanity' (cited in Massey 1991: 160). This moral aversion to the bomb was a strong motivation for Nehru's support for nuclear disarmament. Yet, it was also a political aversion based on a rational assessment of India's national interests. Nuclear weapons represented a threat to world peace. A nuclear war between great powers would have also engulfed most of the globe, including India. Nuclear arms race also engendered severe tensions between the two great powers and therefore complicated India's policy of non-alignment. A stable international system where India could seek help from both great powers was essential to its economic and political development. Lastly, given the shortage of material power, the cause of nuclear disarmament provided Nehru a moral heft in international politics.

The third vector was Nehru's realpolitik instincts in understanding world politics. Moral aversion notwithstanding, his political calculus against nuclear weapons was also inspired by his deep understanding of the importance of power in inter-state politics.[1]

[1] On Nehru's ideas of power in inter-state politics, see Raghavan (2010).

For an independent India, as Nehru had confessed in February 1947, 'defence' was a 'primary need'. It was for this reason that Nehru also saw in the development of atomic energy an option for India's nuclear weapons future. He argued, explaining India's defence policy in February 1947, 'the probable use of atomic energy in warfare is likely to revolutionize all our concepts of war and defence. For the moment we may leave that out of consideration except that it makes it absolutely essential for us to develop the methods of using atomic energy for both civil and military purpose.'[2] This strategy of 'hedging' against the duality of the atom made Nehru invest heavily in research and development (R&D) of atomic energy in India (Narang 2016). Nuclear science, therefore, may help with both development and, if the need be, India's defence. This ambiguity in Nehru's approach towards the atom drove India's nuclear trajectory under his leadership.

If Nehru was the political force behind India's atomic journey, Homi Bhabha was his comrade-in-arms (Deshmukh 2003). During his early research career, Bhabha, a Cambridge-educated scientist, had worked with leading nuclear scientists of the early

[2] See Nehru's note on 'Defence Policy and National Development' of 3 February 1947, reproduced in Singh (1988: 45–6).

atomic era, such as John Cockcroft, Paul Dirac, Neils Bohr, and Wolfgang Pauli. Upon his return to India in 1939, Bhabha took on a professorship in the Indian Institute of Science in Bangalore. By 1940, his work on cosmic rays had earned him the prestigious Fellowship of the Royal Society at a young age of 31. His heart, however, was in atomic research. In 1944, Bhabha established the Tata Institute of Fundamental Research (TIFR) through a grant provided by his uncle and one of India's leading industrial tycoons, Sir Dorabji Tata. The reasoning behind the institute, as Bhabha had explained to J.R.D. Tata, was to create an indigenous scientific manpower which could help India master the nuances of atomic research:

> It is absolutely in the interests of India to have a vigorous school for research in fundamental physics ... [for] ... when nuclear energy has been successfully applied for power production, in say a couple of decades from now, India will not have to look abroad for experts but will find them ready at home. (Andersen 1975: 33)

This was the beginning of India's engagement with atomic science. Bhabha's views on the role of science and technology were similar to Prime Minister Nehru and both accepted that the 'energy potential of the atomic nucleus was the apogee of science' (Perkovich 1999: 17).

Bhabha's scientific credentials, his leadership of atomic energy research in India, his influential background, his belief in the promise of the atomic age, and his personal friendship with Nehru provided him immense influence in shaping the course of India's nuclear future. By 1947, Bhabha had started lobbying Nehru for institutionalization of atomic research in India. This led to the enactment of the Atomic Energy Act of 1948. Through the Act, Nehru and Bhabha aimed at creating an Atomic Energy Commission (AEC) to oversee all aspects of nuclear science and technology research in the country and to provide a legal framework under which the commission can operate. The Act bestowed upon the AEC the sole authority for nuclear research in India. It also labelled atomic research as a state secret, providing unmitigated authority to the AEC over matters concerning atomic R&D. On 10 August 1948, the AEC was established and was directly answerable to Prime Minister Nehru, providing Bhabha unfettered access to the most important political decision-maker of modern India. Even though the political control rested with the prime minister, owing to Bhabha's expertise and his equation with Nehru, functional command lay completely with the chairman of the AEC. This made it one of the most autonomous bureaucratic institutions in Nehru's India. With little political oversight and immense politico-scientific interest in atomic energy, the research in this field grew exponentially, at least in terms of the financial allocation it received from the state.

Yet, till 1954, as the official history of atomic energy in India suggests, no tangible progress was made (Sundaram et al. 1998). The intervening years were largely used for building the resource base for atomic energy research, especially the training of scientists and identification of radioactive material in the country. On the one hand, India had known sources of monazite sands, rich in thorium, but it was not a natural fissile material. On the other hand, India was deficient in the most commonly used source of fissile material, uranium. Whatever sources were found were also of poor quality, with very little content of the fissile isotope, U-235. This resource deficiency, coupled with theoretical possibility of producing artificial fissile materials such as plutonium (Pl-239) and U-233, led Bhabha to conjure a unique nuclear path for India called the three-stage atomic energy programme.[3] The three-stage programme involved a development trajectory which, in the first phase, used natural uranium

---

[3] In November 1954, Bhabha presented this plan in a scientific conference organized in New Delhi and presided over by Prime Minister Nehru. For his presentation, see Bhabha (1956).

(U–238) as the primary fuel for India's nuclear reactors. Irradiation of this natural uranium produced plutonium as spent fuel which, upon reprocessing, could generate a highly fissile isotope called Pl–239. The second stage intended to use Pl–239 as fuel in breeder reactors alongside natural uranium and thorium. Irradiation of thorium in breeder reactors resulted in U–233, which was as good a fissile material as was naturally occurring U–235. Therefore, for the third and the last stage, Bhabha envisioned nuclear reactors based on U–233 as the primary fuel. This three–stage process provided a way out of India's deficient resource base. This was Bhabha's 'grand vision' for atomic energy in India (Deshmukh 2003: 56). By default, it also provided India the necessary wherewithal for a nuclear weapons programme; Pl–239 could also be used for making atomic weapons. Whether the choice of this unique path for atomic energy programme was a camouflage for development of nuclear weapons continues to be debated by historians of India's nuclear programme.

The year 1954 was also a landmark year for Bhabha insofar he further consolidated his grip on the management of atomic energy in India with the establishment of the DAE. The AEC, formed in 1948, was only an advisory body to the Government of India and worked under the Ministry of Natural Resources and Scientific Research. In the intervening years, however, the AEC's activities had expanded dramatically. In 1954, Bhabha convinced Nehru to establish a separate department in the central government to look over matters concerning the atomic energy establishment. The DAE came directly under the PMO and was manned exclusively by nuclear scientists, as was the case with the AEC. This ensured a central role for the DAE and the PMO in India's nuclear decision-making.

In Nehru and Bhabha, India found what the famous Pakistani physicist and Noble Prize winner, Abdus Salam, had once cited as a prerequisite for scientific progress in the developing world: 'the supply of towering individuals, tribal leaders around which great institutions are built' (Salam 1966: 462). Between 1947 and 1954, Nehru and Bhabha had laid the foundations of atomic energy development in India. Their common views on science and technology and the potential of atomic energy in leapfrogging India towards modernity made nuclear research a well-funded scientific enterprise in the country. They had created a unique institutional and bureaucratic arrangement for management of atomic energy in the form of the DAE and the AEC. The Indian state had thus imposed its absolute control over all aspects of atomic research. Yet, till 1954, no tangible progress had been made on producing power out of the atom. India was yet to operate even a research reactor, leave alone the question of producing electricity from nuclear reactors. Being a parliamentary democracy, this slow pace of progress also invited severe criticisms. Therefore, over the next decade or so, the DAE and Bhabha excessively focused on producing tangible results in the field of atomic energy, most importantly in establishing nuclear reactors. This nuclear infrastructure could not have been produced indigenously. For all of Bhabha's insistence on self-reliance, India's initial major breakthroughs in atomic energy came with foreign assistance. This is explained in the next section.

## International Cooperation and the Growth of Atomic Energy

In March 1955, India decided to build its first research reactor, APSARA. This 1 megawatt thermal (MWT) swimming pool-type reactor was to be built with the assistance of the United Kingdom Atomic Energy Authority (UKAEC). Though constructed and designed by Indian scientists and engineers, the UKAEC provided the entire load of enriched uranium fuel. The generosity of the UKAEC is explained by two factors. First, in December 1953, President Eisenhower had announced the Atoms for Peace programme.[4] Under the programme, the US offered to assist other states in peaceful uses of nuclear energy. Being the leader of the Western bloc, this also inspired other technologically advanced states such as the UK and Canada to share their technological progress in nuclear sciences with other countries.[5] The second factor was much more personal. Sir John Cockcroft, the head of the UKAEC, was friends with Homi Bhabha since their Cambridge days in the UK. The construction of the reactor was completed within a year and the reactor was commissioned in August 1956. At that time, India was the first Asian nation to have a nuclear reactor. For Nehru and the AEC, this was a moment of great pride. India had taken a significant lead in nuclear research, which had to be maintained at all costs. International prestige notwithstanding, commissioning of APSARA had a major impact on the research infrastructure which Nehru and Bhabha had so adroitly built since 1947. For the first time, Indian nuclear scientists could

[4] For details on Atoms for Peace programme, see Bader (1968).

[5] A good account of competition among developed countries to supply nuclear technology and materials after the Atoms for Peace programme is available in Walker and Lonnroth (1988).

study and experiment the splitting of the atom in reality; until then, atomic energy research in India had largely been a theoretical enterprise.

APSARA, however, was just the beginning and it did not fit strictly into Bhabha's three-stage nuclear energy programme as it was based on enriched uranium. Even when Bhabha's strategy was to take all possible help from all possible sources, his gaze was fixed on natural uranium–plutonium–thorium fuel cycle he had envisioned in 1954. The specific requirements of the three-stage nuclear programme mandated a reactor design which ran on natural uranium as fuel. In 1950, Canada had initiated a cooperative economic development plan for Commonwealth countries called the Colombo Plan (Oakman 2010). After the Atoms for Peace programme of the US, Canada included assistance in nuclear energy research under the Colombo Plan. For Bhabha, this was a great opportunity because the only two countries that had based their nuclear energy programme on natural uranium were Canada and France.[6] Canada had developed a natural uranium, heavy water-moderated National Research Experimental (NRX) reactor in

[6] Hecht (1998) provides an excellent account of the French nuclear programme. For Canadian nuclear programme, see Pedeo (1976).

the late 1940s. In 1955, when Canada offered India the design and construction of an NRX nuclear reactor with full technical and financial assistance under the Colombo Plan, the offer was 'too good to be ignored' (Sundaram et al. 1998: 21).[7] Again, Bhabha's personal friendship with W.B. Lewis, the head of the Canadian Atomic Energy Agency, was instrumental in the smooth negotiations of the deal. In 1956, India and Canada signed an agreement for construction of an NRX type of reactor at Trombay. Subsequently, India entered into an agreement with the United States Atomic Energy Commission (USAEC) for purchase of heavy water to be used as the moderator in this reactor (Wit and Cubock 1958). Canada also offered to supply the entire load of natural uranium fuel, but Bhabha decided that the DAE will build half of the initial fuel load. A number of Indian scientists and engineers were also trained at the Canadian nuclear facility at Chalk River. The Canada–India Reactor (CIR) project was completed in early 1960 and the reactor, christened CIRUS (Canada India Reactor Utility Service), achieved criticality on 10 July 1960. It represented a giant leap for the Indian atomic energy establishment: at 40 MWT, it was the most powerful reactor in

[7] An alternate account of India–Canada peaceful nuclear cooperation is available in Srinivasan (2003).

Asia at the time. Most importantly, CIRUS was the foundation of India's three-stage nuclear programme. It was a prototype of India's most successful nuclear power reactor design, the Canada Deutarium Uranium (CANDU) type of reactors. It was also India's first source of plutonium.

The DAE began work on a purely indigenous research reactor soon after the construction of the CIRUS. This zero-energy critical reactor was called ZERLINA and attained criticality in January 1961. Its most important contribution was in the investigation of fuel assembly behaviour under nuclear reactions. This indigenous moment in nuclear energy development took a great leap forward with India's first plutonium reprocessing plant. Even though CIRUS had provided India its first major breakthrough in implementing the three-stage nuclear programme, central to Bhabha's grand vision was plutonium reprocessing. The spent fuel from the natural uranium-based reactors required further chemical treatment before highly fissile Pl-239 could be extracted out of it. Known as reprocessing, this was a very sophisticated and technologically challenging scientific process. Bhabha, therefore, paid special attention to plutonium reprocessing. As work on CIRUS was progressing, India decided to build a plutonium reprocessing plant at Trombay in July 1958. Called Project Phoenix, it was entirely a handiwork of

Indian scientists and engineers and was led by Homi Sethna, who later became the chairman of the AEC. Designed to reprocess around 20 tonnes of spent fuel a year, it had a capacity to produce 10 kilogram (kg) of Pl-239 annually (Sundaram et al. 1998: 90–1). When it was completed in mid-1964, India was the fifth country after the US, the UK, France, and the Soviet Union to have such a capability.

Within a decade of laying down his grand vision for atomic energy, Bhabha was able to put in place some of the most critical components of India's atomic energy programme. Yet, even with APSARA, CIRUS, ZERLINA, and Phoenix, India had not produced an iota of nuclear energy. This was indeed the most damning side of India's atomic programme as nuclear energy, both in Nehru and Bhabha's articulation, was the principle reason behind their pursuance of atomic science. Foreign assistance was therefore sought. In 1962, General Electric (GE) Company of the US was chosen to build two power reactors at a place called Tarapur. The GE reactors were light water moderated and fuelled by enriched uranium and did not fit the three-stage atomic energy programme. Construction work on Tarapur nuclear power plants began in October 1964 and was finished in 1968. On 28 November 1969, the Tarapur Atomic Power Stations (TAPS) started producing electricity.

Between 1955 and 1965, Nehru and Bhabha's vision of an atomic future for India made major strides. By the time of Nehru's death in May 1964, India was considered one of the most advanced nuclear technology states in the Third World. India's capabilities in nuclear science and technology also brought her closer to mastering not only the peaceful uses of nuclear energy but also its destructive potential. As we will see in the next chapter, by 1964, India was considered to be on the threshold of acquiring a nuclear weapons capability. However, the most distinctive feature of this period of growth in atomic energy in India was its international dimension. For all the focus on self-reliance, India's first major achievements in atomic energy were a result of international cooperation and assistance. Henceforth, technical cooperation in peaceful uses of nuclear energy remained a fundamental aspect of India's nuclear policy. It also translated into India's aversion to strict control of nuclear technology and material by the advanced nuclear technology states.

## International Control over Nuclear Technology

From the very beginning, Nehru and Bhabha's atomic vision for India collided with the urge of advanced nuclear states to establish some kind of control over nuclear technology and material.[8] The battle was between two principles. On the one hand, Third World countries like India saw in nuclear energy a means to make rapid advancement towards economic development, and therefore sought assistance from the advanced technology states. On the other hand, nuclear powers like the US saw in dispersal of nuclear know-how a venue for proliferation of nuclear weapons (Haskins 1946). This not only threatened international peace but in the immediate aftermath of the Second World War when the US was the sole nuclear power in the world, it also posed a challenge to its nuclear dominance. Therefore, in 1946, the US government proposed a plan to create an International Atomic Development Authority (IADA), also called the Baruch Plan. Under the plan, the IADA was to be given the authority to control, own, and operate all nuclear technologies and materials, including those which can have potential weapons uses. The Baruch Plan was originally targeted at Soviet Union's efforts to build a nuclear bomb, at least so was the Soviet perception (Goldsmith 1986). If implemented, it would have had implications for India as well. Its thorium and uranium deposits would have come under international control.

[8] For a history of international controls over nuclear energy, see Goldschmidt (1982).

For a country which had recently gained independence, maintaining its sovereignty was a major concern (Bhatia 1979: 43–4). India, therefore, resisted the US plans for international control of nuclear technology and materials. Resistance from other states was equally fierce and with the Soviet Union going nuclear in 1949, the plan suffered a quiet burial.[9]

A much-diluted version of the plan was, however, announced by President Eisenhower in December 1953 during his Atoms for Peace speech at the United Nations General Assembly (UNGA). The Atoms for Peace Plan envisaged sharing nuclear material and technology for peaceful purposes, but to restrict proliferation of nuclear weapons by mandated oversight over other's nuclear programmes through an international agency. India could visualize several problems regarding this plan. First, not all states were members of the UN and in some cases, such as China, would not agree to any controls over their nuclear programmes. Second, international control was synonymous with colonialism and given the asymmetry in nuclear knowledge between countries with advanced technological know-how and the Third World, the recipient states would always be dominated

---

[9] For a historical account of Soviet nuclear weapons programme, see Holloway (1994).

by the major nuclear powers. Supplier states, therefore, would write the rules of the game.

By 1955, however, there emerged a general agreement among major nuclear powers to establish the IAEA. During the First Geneva Conference on Peaceful Uses of Nuclear Energy in 1955, Homi Bhabha was unanimously chosen as its chairman. As India was one of the most advanced countries in nuclear science and technology in the developing world, its participation in the negotiations was vital for the effectiveness and legitimacy of the agency. It also provided India a chance to shape the IAEA in a way that best suited its own requirements. The most important agenda for India in the IAEA negotiations was the issue of safeguards.[10] If for the advanced nuclear states the bargain had to be between sharing their nuclear technology and materials and the willingness of the recipient states to give up the option of developing nuclear weapons, it was the IAEA which had to oversee that this bargain was followed in letter and spirit. The IAEA must, therefore, have the power to maintain safeguards over technology and materials provided by advanced technology states. The problem for Bhabha

[10] A good summary of India's attitude towards international controls on atomic energy can be found in Sullivan III (1970).

was whether the plutonium produced using foreign reactors such as CIRUS would come under the control of the IAEA. Even though CIRUS was negotiated before the IAEA came into being and only bilateral safeguards were applicable to it, future negotiations over foreign assistance would have come under the purview of the IAEA (Perkovich 1999: 27–8). This would have severely curtailed India's three-stage nuclear energy programme. India, therefore, vociferously opposed the strict safeguards regime as envisioned during the IAEA draft statute negotiations in 1955 and 1956. In the end, the final statute of the IAEA declared a much-relaxed safeguards regime where recipient countries could hold onto the plutonium extracted through spent fuel generated in foreign-supplied nuclear reactors and use it for peaceful purposes.

Till the time the IAEA came into being in 1957, all of India's nuclear agreements with foreign countries such as the UK, Canada, and the US followed strictly bilateral safeguards mechanisms (Greenberg 1962). Bhabha was able to get generous treatment from these foreign collaborators for a number of reasons. First was their willingness to provide nuclear technology and materials because non-proliferation had not emerged as a major concern till then. There was also a competition among advanced nuclear technology states to sell their nuclear know-how and establish themselves as major nuclear exporters. Second factor was Bhabha's interpersonal relationships with scientists from these countries, as was the case with John Cockcroft of the UKAEC and W.B. Lewis of the Canadian Atomic Energy Agency. Lastly, Bhabha's scientific stature and a global perception of India's stature in nuclear research and being a leading Third World country further helped its cause. However, India's discomfiture with international controls over nuclear technology and materials was evident from the time the US had proposed the Baruch Plan in the late 1940s. India's objective of maintaining its strategic autonomy on the development of its nuclear programme often collided with the urge among major nuclear powers to restrict proliferation of nuclear weapons by insisting on tougher safeguards.

## India's Diplomatic Leadership in Nuclear Disarmament

If India's nuclear pioneers were adamant in preserving the country's atomic sovereignty from international efforts to control nuclear know-how, they were equally emphatic in propagating India's diplomatic leadership in preserving international peace from the fallouts of an unmitigated nuclear arms race. There was an element of moral repugnance towards nuclear weapons which was prevalent in Indian political thought. Nehru's enthusiasm for nuclear disarmament was also driven by pragmatic considerations. For one, emphasizing nuclear disarmament to both power blocs increased India's credibility as a genuine non-aligned state. Second, it allowed India to play a much more significant role in global nuclear diplomacy than was warranted by its low material power. Third, it also helped increase India's credibility and stature among the developing countries. Lastly, unlike the issue of international control of atomic energy, India had no direct stakes involved in nuclear disarmament since it was far away from having a nuclear weapons capability. Its leadership, therefore, came with no obligation.

These considerations explain India's disarmament diplomacy under Nehru's prime ministership. It was underlined by three principles (Sharma 2013). First was India's insistence that international bodies, such as the United Nations Disarmament Commission (UNDC), discussing arms control and disarmament must not exclude non-nuclear weapon states (NNWS). Left to their own devices, nuclear weapon states had little incentive to disarm. Multilateralism in disarmament negotiations, therefore, became one of the cornerstones of India's disarmament diplomacy. Second, even when India sought greater representation, it strictly maintained a position that the primary obligation for nuclear disarmament was of the great powers. Third, the complexity of nuclear disarmament required a step-by-step approach. Radical solutions to nuclear disarmament were highly impractical because nuclear weapons had now become integral to the national security of nuclear weapon states.

India's first major disarmament initiative in 1954 was a combination of these principles. The Soviet nuclear test in 1949 had unfurled a test race between the US and the Soviet Union. This race was not only focused on the number of tests but on their increasing destructive firepower, helping the two sides build nuclear weapons of higher and higher yields. By 1952, the US had introduced hydrogen bomb into the Cold War nuclear dynamics. With the Soviet Union's thermonuclear weapon tests in 1954, both sides now mastered the ultimate weapon in human history. Nuclear tests not only fuelled an intense arms race but were also deleterious for earth's environment. It was therefore that, in April 1954, Nehru proposed in the Indian Parliament a cessation of further nuclear tests by the US and the Soviet Union calling for a 'standstill agreement' (Jayaprakash 2000). This was India's first major disarmament plan.

The 'standstill agreement' Nehru proposed was, however, just one part of his disarmament initiative. Other parts included making the world aware of the destructive potential of the hydrogen weapons so as to generate global public opinion against these weapons; immediate consultations in the UNDC to negotiate the cessation of further testing; and rallying non-nuclear states to bring their influence to bear upon the US and the Soviet Union to stop further development of their nuclear arsenals. Nehru's 1954 appeal encapsulated all three principles of India's nuclear disarmament diplomacy. By involving non-nuclear states in negotiations of the 'standstill agreement', it sought to pursue disarmament multilaterally. Yet, the obligation to cease nuclear testing was primarily of the US and the Soviet Union. Lastly, it was in consonance with a step-by-step approach to nuclear disarmament and did not intend to eliminate nuclear weapons in their entirety at one go.

The 'standstill agreement' became the lynchpin of India's disarmament diplomacy in the 1950s.[11] Even when it was largely ignored by the superpowers, it did earn India considerable diplomatic clout in the UN and among the Third World nations. It established India's moral and diplomatic leadership in the field of global nuclear disarmament. Its major achievement, however,

[11] For India's efforts on nuclear disarmament in the UN during the 1950s and 1960s, see Reddy and Damodaran (1994).

was that it put the need for a nuclear test ban firmly on the global disarmament agenda. The quest for ban on nuclear tests gained traction in 1958 when Soviet Union announced a unilateral moratorium. The first major global disarmament treaty that came into effect in 1963 was the Limited Test Ban Treaty, proscribing nuclear tests on land, sea, and air, except those carried underground. India's role in this breakthrough was, first, ideational. It was Nehru who was both the originator of the idea behind a test ban and also its most emphatic public face. Second, even when it was rejected by the superpowers, India maintained a diplomatic crusade in the UN and other forums for the need to stop nuclear tests.

India's first major engagement on nuclear disarmament thus happened in a period when it had no major stakes in nuclear disarmament. Even when rapidly advancing in its atomic energy capabilities, it was still a non-nuclear weapons power. Moreover, during this period, it also had no major motivation to go nuclear as it remained unthreatened by a hostile nuclear power even though, as we saw earlier, Nehru had adopted a strategy of nuclear hedging. With the Chinese nuclear test in October 1964, India's national security requirements altered substantially. Henceforth, national security, not moral or diplomatic influence, would become the benchmark for its engagement with nuclear disarmament and arms control. Yet, the rhetoric of nuclear disarmament would also be the pretext on which India would reject participation in any arms control measures. In later decades, disarmament helped India practice realism under the cloak of idealism.

★★★

India's nuclear pioneers, from the very beginning, understood the significance of atomic energy. This appreciation grew out of their technological vision for modern India and also because of the promise inherent in the atom. On the one hand, atomic science promised a rapid march towards economic and social development when used for harnessing energy. On the other hand, it also provided a destructive power of immense consequences in the form of nuclear weapons. Atomic energy was also a source of international prestige as Nehru sought India's leadership of the non-aligned and the developing world. These beliefs principally shaped the evolution of atomic energy research in India. In the process, Nehru and Bhabha not only laid the foundations of India's atomic energy programme but also ensured that it received the maximum patronage among all other scientific endeavours in independent India. If, for Nehru, science and technology had to create the 'temples of modern India', atomic energy became the holiest of these shrines. The atomic energy establishment therefore grew and by 1964, India was one of the most advanced states in nuclear technology in the entire Third World. For all the emphasis on self-reliance, however, easy access to foreign nuclear technology and materials, especially from the UK, the US, and Canada, was critical to India's nuclear progress. Yet, India doggedly preserved its strategic autonomy and declined to accept any international controls on its atomic energy programme. Nehru's disarmament diplomacy only expanded India's influence and its leadership on nuclear issues. In all, the direction and momentum provided by India's nuclear pioneers to its atomic energy programme allowed India to become the loadstar of atomic research in Asia and among the Third World.

# 2

# Perils of a Nuclear Neighbour

Till 16 December 1964, when China exploded a nuclear
device at Lop Nor, India had not yet faced a direct
nuclear threat. Nuclear weapons did pose a problem for
Indian foreign policy but in an oblique sense. First, given
the idealist non-violent underpinnings of India's struggle
for independence and its post-independence foreign
policy, there was a moral abhorrence to such weapons
of mass destruction.[1] Such moralistic considerations
notwithstanding, the nuclear arms race between the
two superpowers was harmful insofar it led to intense
security competition and destabilized the international
system. A stable international system was important for
India's internal development, and it was equally critical
for the policy of non-alignment as tensions between the

[1] The idealist streak in Indian foreign policy is best
explained in Brecher (1968).

two superpowers made it difficult for India to chart a middle course in its foreign policy. Nuclear arms races were also a principal source of international tension as was the case during the Cuban crisis of 1962. However, as New Delhi had learnt during the Sino-Indian war in late 1962, the spectre of superpower confrontation allowed countries like China to pursue their revisionist aims against their smaller neighbours (Raghavan 2013).

China's nuclear test was, however, a radical departure. For the first time in independent India's history, a hostile neighbour, with which it had fought and lost a brief but intense border war in 1962, had acquired a nuclear capability. The conflict remained alive with China claiming huge swathe of Indian territory along the Himalayan frontier. Second, if India's territorial integrity was threatened, so was its international prestige. China's bomb, as an Indian diplomat had argued then, was a crisis in 'India's manifest destiny' to be a major global power: if nuclear technology represented a domain of technological progress, China had left India behind.[2] This threatened New Delhi's purported

---

[2] In November 1964, K.R. Narayanan was the Director of the China Division in the Ministry of External Affairs (MEA). As a response to the Chinese nuclear test, he wrote a 10-page policy note calling for an Indian nuclear weapons programme, see NAI (1964).

leadership of Asia and of the Global South. If becoming a global power was always a major goal in India's post-independence foreign policy, China had taken a great leap forward over India in the struggle for global status. Third, nuclear weapons in the hands of a Maoist China allowed her to aggressively export its revolutionary ideology to India's body politic; Beijing had hailed it as a moral boost to wars of national liberation (Halperin 1965). The nascent Indian democracy was struggling with communist revolutions, such as the Naxalbari movement, and revanchist insurgencies in its north-east. Chinese bomb was both an exemplar and enabler to such anti-state ideologies (Doctor 1971). Finally, this was also the period of internal flux in India. Prime Minister Nehru's death in May 1964 had not only left a void in internal leadership of the Congress but had also resulted in the loss of India's most important source of international influence. The question staring at India's post-Nehru leadership, as one commentator argued then, was whether 'India should exercise the option of becoming a nuclear power, a path which many other countries have insisted leads to national unity, national security and international prestige' (Buchan 1965: 210).

The events of October 1964, for the first time, brought the issue of nuclear weapons to the forefront of India's security policy. This chapter charts the course of India's nuclear behaviour and policy in the crucial decade following China's nuclear test. It parses the debate within, India's responses, and the failure and success of its nuclear policies between 1964 and 1974. Rather than undertaking a full-fledged nuclear weapons programme, India primarily depended upon diplomacy to counter the challenge posed by China's nuclear capability. These diplomatic strategies included seeking nuclear security guarantees from established nuclear powers, such as the US, the Union of Soviet Socialist Republics (USSR), and the UK. The NPT was equally an instrument in India's diplomatic arsenal. However, New Delhi also hedged towards a 'nuclear option' by initiating technical and scientific development towards a nuclear explosion capability. Yet, India consistently maintained a public aversion to nuclear weapons, often shrouding its nascent nuclear capabilities under the rubric of 'peaceful nuclear activities'. Even when China had given India a reason to go nuclear, New Delhi's nuclear policy remained highly ambivalent.

## India's Quest for a Diplomatic Deterrent

In the aftermath of the Chinese tests, the central question for the Indian leadership was to think of a response to the threat posed to Indian security by China's possession of the bomb. Notwithstanding the cacophony of the public debate on exercising India's nuclear option, the political leadership headed by Prime Minister Shastri chose a diplomatic strategy to deter China (Kennedy 2011). Shastri sought security guarantees from the dominant nuclear powers: the US, the UK, and the Soviet Union. Extended deterrence where dominant nuclear powers provided the cover of their nuclear arsenals to protect their allies from nuclear threats was by then a norm in Cold War dynamics. The problem, however, remained one of credibility: would established nuclear powers risk self-destruction for the sake of their lesser allies (Coffey 1971). The problem was compounded by the fact that India was a non-aligned country rather than an ally of any of the superpowers and maintained a policy of non-alignment. These contradictions eventually proved too difficult for India to surmount and led to the failure of this diplomatic strategy. India wanted a nuclear umbrella but one 'without a handle' (Noorani 1967: 500).

India's quest for security guarantees was first inspired by US President Lyndon Johnson's speech just two days after the Chinese test. On 18 October, Johnson had publicly offered 'to respond to requests from the Asian nations to help in dealing with Communist China's aggression' (cited in Gavin 2012: 93). Though the promise was vague, it did outline a possibility where states threatened by Chinese nuclear bomb could chalk out the specifics of a US nuclear guarantee. The initial Indian response was lukewarm but as the domestic pressure for an indigenous nuclear weapons programme mounted, Prime Minister Shastri saw in it a way out from the difficult decision to undertake a full-fledged nuclear weapons programme. He, therefore, appealed for a nuclear umbrella during a meeting with British Prime Minister Harold Wilson in December 1964 (Schrafstetter 2002). Furthermore, his decision was also reflective of his inexperience with nuclear policy. Under Nehru, India's nuclear decision-making was highly concentrated in the PMO and none of the other leaders in the Indian National Congress were conversant with the issue. Prime Minister Shastri did not really consult anyone in the Indian cabinet, or in the civilian bureaucracy, on the issue of nuclear guarantees (Brecher 1966: 127).

Notwithstanding the reasons behind Shastri's appeal, his framing of a security guarantee balanced a number of competing interests. India needed some kind of protection from a nuclear blackmail or use of nuclear weapons against it. Yet, any India-specific proposal would have been an acceptance of weakness on the part of the political leadership. Moreover, asking all 'nuclear powers' to guarantee protection allowed Shastri to avoid being criticized for abandoning the basic precepts of non-alignment. These competing interests refused to be reconciled, however. For one, the nuclear guarantors—the US and the UK—found it difficult to credibly associate themselves with guaranteeing India's security, especially when it was not a treaty ally, even though India's security may have been 'vital for the security of the [Asian] region as a whole' (Coffey 1971: 839). Soviet reluctance, on the other hand, was engendered due to China being a fellow communist power, and also in the hope that a muted reaction would lead to a rapprochement in Sino-Soviet relations.[3] Domestically, the Indian government was attacked by both the votaries of non-alignment and the advocates of the indigenous nuclear weapons programme. The former saw in the request for nuclear guarantees a backdoor entry into the power politics of the Cold War.[4] The latter decried it as a policy of escapism.[5] Soon after, the Indian government clarified

[3] A good account of Soviet nuclear policy between the Chinese nuclear test and the NPT is available in Larson (1969).

[4] These views were most emphatically argued by Krishna Menon. See Brecher (1968).

[5] These were largely made by the pro-bomb lobby represented by the right-wing political forces. A good summary of their views is available in Erdman (1967) and Kishore (1969).

that any such guarantee had to be jointly managed by the two superpowers: the US and the Soviet Union. Doing so allowed India to cover its request for superpower intervention under a 'non-aligned cloak' (Edwardes 1965: 57). However, the response from the nuclear powers remained ambivalent. For the West, Johnson's vague promise made in October 1964 sufficed. Any further refinement of the promise had to be requested by India, which given the long-held Congress party's position against military pacts was difficult to come by. The Soviet Union, meanwhile, had different considerations. On the one hand, it did not seek to further rattle the Chinese and on the other, Moscow was wary of the security guarantees because it would have allowed the West to station its nuclear assets, especially its ballistic nuclear submarines, in the Indian Ocean.

Against this background, it was highly doubtful that the established nuclear powers would offer any positive security guarantees. New Delhi's diplomatic strategy to counter the Chinese nuclear weapons therefore shifted to the NPT. The Chinese nuclear test had put the issue of proliferation of nuclear weapons at the forefront of international security. In May 1965, the 114-member UNDC debated the issue of the NPT. India saw in it a method to restrict further advancement of the Chinese nuclear capability as well as a way to secure the nuclear guarantees from the nuclear weapon states. In its submission to the UNDC, India laid down five conditions for New Delhi to join the treaty (Perkovich 1999: 103):

1. Nuclear powers should not transfer nuclear weapons or nuclear weapons technology to others.
2. Nuclear powers should agree not to use nuclear weapons against non-nuclear states.
3. United Nations must guarantee the security of those countries which are threatened by 'nuclear weapons or states near to possessing nuclear weapons'.
4. Tangible progress on nuclear disarmament should be made, including a CTBT, freeze on production of nuclear weapons and their means of delivery, as well as substantial reduction on existing stockpiles.
5. Non-nuclear powers should not acquire nuclear weapons.

These conditions, if incorporated in the treaty, could have allayed India's concerns as it would have stopped the development of Chinese nuclear programme in its tracks. Also, if China declined to be a part of such a treaty, security guarantees recommended under the treaty would have created a deterrent against any Chinese aggression or nuclear blackmail. More importantly, it was a policy to escape criticism at home both from those who were lobbying for an indigenous nuclear programme and those who saw in India's quest for security guarantees a dilution of non-alignment. Multilateral nuclear security guarantees under the UN would have allowed India to maintain the policy of equidistance from the superpowers while enjoying a deterrent against China.

Indian hopes met stiff resistance from the established nuclear powers. In June 1965, the UNDC passed a US-moved resolution to discuss the NPT at the Eighteen Nation Disarmament Committee (ENDC) in Geneva. When discussion started at the ENDC in early 1966, nuclear powers were ready to offer only negative security guarantee: to agree not to use nuclear weapons against non-nuclear states (Noorani 1967: 495). In other words, nuclear powers were disinclined to positively secure the interests of non-nuclear states by extending their nuclear deterrence to them. Through the NPT, India sought a multilateral nuclear guarantee under the UN, as bilateral guarantees would have amounted to a military pact and seriously punctured its policy of non-alignment. Multilateral guarantees, in other words, were just not available.

Even when the quest for security guarantees was discarded by the nuclear haves, India still went ahead with negotiating the NPT at the ENDC. India decoupled its participation in the NPT process with the positive security guarantees it had asked in May 1964. Rather India now focused on nuclear disarmament. It, therefore, asked nuclear powers to make concrete promises on the elimination of nuclear weapons under the NPT. For one, even when China had conducted another nuclear test in the summer of 1965, India still believed that concrete commitments on nuclear disarmament could arrest China's march towards a sophisticated nuclear arsenal with advanced delivery systems. There were also other reasons to support the NPT. First, the NPT process allowed India to isolate China internationally. If the Chinese nuclear bomb had led to China's considerable influence among the Third World, India was to regain its leadership role through its quest for nuclear disarmament. Second, the NPT process, insofar it represented the process of détente between the great powers, helped India's cause. A sense of stability in the superpower relations could deter China from exploiting tense relations between the US and the Soviet Union for its own ends in Asia. Lastly, continued participation in the NPT negotiations dispelled global fears that India was mulling a nuclear deterrent of its own. Yet, India was adamant in keeping its option open by supporting peaceful uses of nuclear energy, including PNEs.

As the NPT negotiations unfolded in 1966, the divergence between India and the other nuclear powers continued to grow. The nuclear powers did not pay heed to any of India's concerns: elimination of further production of nuclear weapons and delivery systems; commitments on nuclear disarmament; on providing security guarantees; and on right to use nuclear energy for peaceful purposes. They rather focused exclusively on the proliferation side: preventing nuclear weapon states from transferring nuclear weapons and technology to non-nuclear states and make the non-nuclear states give up the right to pursue nuclear weapons. India was not ready to accept such one-sided commitments because it both violated the principles of equity and hardly addressed its security concerns.

India, however, did make one last effort to synchronize its concerns with the NPT. In spring of 1967, Prime Minister Indira Gandhi dispatched a special envoy, L.K. Jha, to Moscow, London, and Washington, DC, to seek clarifications on how the established nuclear powers could protect non-nuclear states from blackmail or use of nuclear weapons. The conclusion of Jha's mission, now available in the form of top-secret memos in the Indian archives, was categorical: even though the nuclear powers such as the US, the UK, and the Soviet Union were sympathetic to India's concerns and would intervene in the event of hostilities with China, no formal nuclear guarantee would be available.[6] As Jha wrote to Prime Minister Gandhi in May 1967, 'a political guarantee is possible, but a legal guarantee is impossible.' For Jha, 'since neither the USA nor the USSR can afford to let India go under Chinese domination,' their 'political compulsions' offered a political guarantee against any Chinese nuclear threat or blackmail.

The nuclear powers, however, would not agree to a formal treaty with 'commitments to help any country at any time irrespective of the circumstances'. Therefore, he argued that India should not let go the option of developing nuclear weapons entirely. Jha recommended to the prime minister that NPT should only be seen in terms of a temporary arrangement towards the ultimate goal of nuclear disarmament. The 'objection in principle', as L.K. Jha put it, was whether 'we and, therefore, other nations too, should continue to have the right to make nuclear weapons as long as any country in the world has the right to do so'. The objective was clear: if the need arises for an indigenous nuclear capability, India should not be tied down by the requirements of the NPT. Therefore, when in

---

[6] Jha wrote two policy memos upon his return: 'Nuclear Policy' and 'Nuclear Security'. See NMML (1967a, b). For a detailed explanation on L.K. Jha's memos, see Joshi (2015).

August 1967 the US and USSR submitted a joint draft of the NPT to the ENDC, India 'deemed the draft inadequate' (Perkovich 1999: 139). In October 1967, Indian Defence Minister Swaran Singh told the UNGA that India will not sign the NPT.

By early 1968, as the draft treaty was debated and discussed in the ENDC, the Indian PMO took a final cognizance of the matter. Top-secret assessments made in the PMO reveal that India's decision not to be a party to the treaty was by now a firm policy decision. The fact that China had declined to be a part of the treaty was the final nail in the coffin of India being a signatory to the NPT. Yet, India did not intend to scuttle the NPT since it could have angered both the US and the USSR. India required both these superpowers to come to her aid in case of a Chinese nuclear threat. However, in India's security policy, even when the superpowers had refused to provide explicit security guarantees, the unspoken assumption was that India was too important to be left to its own devices if push came to shove against China (Babu 1968). The NPT was also a manifestation of the process of détente. Stability in US–USSR relations was important for India as tensions between the two could have been exploited by China to further its revisionist interests in Asia, as was the case after the Cuban missile crisis. Instructions to India's representative to the ENDC from Prime

Minister Indira Gandhi's office were reflective of this balanced approach: 'avoid polemical tone against the nuclear powers'; mention the Chinese threat but 'we should neither overplay that threat nor underplay it'; 'should not mention Pakistan'; 'mention that our policy as hitherto continues to be to refrain from doing anything which would escalate the nuclear arms race'; and 'on the question of the time table for conclusion of the Non-proliferation treaty, we should not spearhead any move for delay and postponement'.[7] India's final act on the NPT was guided by both the need to keep its nuclear option open and the need to maintain an international political environment where China could be isolated.

By 1968, India's diplomatic crusade against China's nuclear capability had come to naught. India's quest for nuclear security guarantees was compromised by its policy of non-alignment on one hand, and the reluctance of the established nuclear powers to offer formal security guarantees to the non-aligned on the other. India had entered the NPT negotiations with a view that the treaty would help in arresting China's nuclear trajectory. However, towards the end, India became a prime target of the treaty with China being recognized as a nuclear weapon power with a right

[7] NMML (1968).

to possess a nuclear weapons programme. As the NPT offered no firm security benefits to New Delhi and created an unequal division between nuclear weapon states and non-nuclear states, India rejected the treaty both on grounds of security and equity. The failure of India's diplomatic enterprise also heralded the first foray into exercising its 'nuclear option'.

## Developing India's 'Nuclear Option'

On 4 October 1964, even before the Chinese nuclear test, Homi Bhabha, the architect of India's nuclear programme, had publicly declared that 'India could explode the bomb within 18 months of a decision to do so' (*National Herald* 1964). Though Bhabha acknowledged that 'no such decision' had yet been taken, few would have doubted his words. In the early 1960s, India was the most advanced state in Asia with a huge nuclear science and technology infrastructure, mostly because of Nehru's emphasis on developing nuclear energy. India's extensive nuclear profile was responsible for a global perception that it was one of the front runners in the race towards a nuclear weapons capability, if the Indian decision-makers decided to have one.

Yet, as we have seen in the previous section, Prime Minister Shastri opted for a diplomatic strategy to counter the Chinese threat. Shastri, however, was under pressure from the pro-bomb lobby in the Parliament, as well as Homi Bhabha, to initiate a nuclear explosive programme. In the aftermath of the Chinese test, Bhabha was actively lobbying for a nuclear weapons programme.[8] He also took on those who criticized an indigenous nuclear capability on the pretext that it was a highly costly affair. For Bhabha, a 10 kiloton nuclear device would have led to an expenditure of merely Rs 17.5 lakh (1.75 million); an inventory of 50 atomic bombs could be managed only at Rs 10 crore (100 million).[9] Bhabha's public sponsorship of the bomb, however, did not sit comfortably with Prime Minister Shastri. For one, Shastri was not ready to sanction a nuclear weapons programme because this would have been a major change in India's traditional position. Being a Gandhian, Shastri also exhibited a moral aversion to the bomb. Also, as prime minister, Shastri knew far

[8] For the domestic clamour for the bomb and slow drift towards proliferation, see Sarkar (2015).

[9] Bhabha's estimates can be found in a study compiled by the Department of Atomic Energy on various estimates of the costs associated in acquisition of nuclear weapons by India. This study also provides figures calculated by experts such as Alastair Buchan, Leonard Baton, James Schlesinger, Subramanian Swamy, and Ralph E. Lapp. See Department of Atomic Energy (1970: 64–5).

better the dire economic health of the Indian state. After the 1962 Indo–China war, the cost of conventional defence had skyrocketed: by 1965, India's defence expenditure amounted to around 4.2 per cent of the gross national product (United States Arms Control and Disarmament Agency 1968). Internally, the country was suffering the worst food crisis in history. These pressures notwithstanding, Shastri was also doubtful of Bhabha's claims. He not only disputed them openly in the All India Congress Committee meeting in early November 1964 but also sought British assistance in providing what he considered to be a more objective assessment of the costs involved in developing a nuclear deterrent.[10]

[10] In December 1964, Indian Prime Minister Lal Bahadur Shastri had requested the Wilson government to offer an analysis of the cost of the bomb in case India decides to develop it. This resulted in a report prepared by British Ministry of Defence titled, *Indications of the Cost of an Indian Defence Capability in the Light of British Experience*. The report suggested that financial implications of the bomb and acquisition of a bomber force to deliver it would be exorbitant: something to the tune of $350 million, with a running cost of $50 million per annum. However, as Schrafstetter (2002: 93–4) has argued, these figures were highly inflated. In a confidential report which was not shared with the Indian government, the actual costs were assumed to be significantly less.

This tussle between the prime minister and India's most acclaimed nuclear scientist led to a necessary compromise by the end of November 1964 (Perkovich 1999: 82). Under this arrangement, Shastri agreed to Bhabha's demands to explore the possibilities of a nuclear explosion; yet it had to only be a theoretical enterprise. Given the government's public disavowal of a nuclear weapons programme and the global concerns over proliferation, Shastri declared in the Parliament that India will explore nuclear explosion technology only for peaceful purposes. Thus was born India's policy of developing a 'nuclear option' (Rajan 1975: 300).

If Bhabha was emphatic in declaring India's nuclear capabilities in public, in private he acknowledged India's constraints. By early 1965, as a top-secret correspondence now available in the Indian archives suggests, Bhabha had revised his earlier estimate of the time period to produce a nuclear explosion from 18 months to at least 'five years'.[11] Major problem for India's scientific community was the issue of fissile material. In January 1965, the Phoenix plutonium separation plant was commissioned. Yet, it faced enormous operational problems. Bhabha, therefore, first explored the possibility of acquiring a nuclear

[11] NAI (1965).

device from the US. In early 1965, the USAEC was keen on exporting nuclear explosive capability for peaceful purposes under what was called the Plowshare programme. Between late 1964 and mid-1965, Bhabha explored this option during his multiple meetings with USAEC officials, especially regarding transfer of plutonium. However, this was also the time when the US non-proliferation policy was taking a concrete shape. As the Johnson administration rallied behind the NPT, the option of acquiring the necessary technology from the US gradually closed. India's nuclear scientists also lacked other necessary expertise for developing a plutonium-based nuclear device.

By early 1966, the 'nuclear option' strategy was hit by the sudden death of both Prime Minister Shastri and Bhabha. Within a span of a month in January 1966, India had lost both its political as well as scientific leadership. If Indira Gandhi was busy consolidating her position in the fractious Congress party during her first year in power, Vikram Sarabhai—Bhabha's replacement at the DAE—openly argued against a nuclear weapons programme and was unsympathetic to the idea of a PNE.[12] Unlike Bhabha, Sarabhai

---

[12] Sarabhai first articulated his ideas on nuclear weapons in a press conference on 1 June 1966. Excerpts of this conference are available in DAE (1970).

argued that a nuclear weapons programme would seriously jeopardize India's economic development as it adversely impacted the 'utilization of national resources for productive and social welfare against the burden of defence expenditure'. Moreover, the problem of India's security as nuclear deterrence could not be achieved by merely 'exploding a bomb'. For Sarabhai, it required a 'total defence system, a means of delivery ... long range missile ... radars ... high state of electronic and a high state of metallurgical and industrial base.' The enormity of this enterprise would have required a 'total commitment of national resources of a most stupendous magnitude' and would have seriously impacted India's economic and social development. Sarabhai, therefore, put an end to the theoretical enterprise of developing PNE technology within Bhabha Atomic Research Centre (BARC), at least formally.

However, as India met with diplomatic disappointments in securing nuclear guarantees and the NPT declined to address India's security concerns, the scientists at BARC began working on the nuclear device in early 1968. It is not clear whether this was expressly permitted by the prime minister but, according to a top-secret document, Homi Sethna, who was heading BARC during this time, had expressly asked for the prime minister's permission to finish the work undertaken under the Shastri–Bhabha compromise in July 1967.[13] Given that the scientists were suffering from 'extreme uncertainty' on this question, P.N. Haksar, secretary to the prime minister, advised Gandhi to take a decision on whether the preparations should continue or not. Diplomatic disappointments and pressure from the nuclear scientists may have led Gandhi to allow the 'most concerted effort yet to develop nuclear explosives' in the Indian case' (Perkovich 1999: 139).'[14] If the availability of fissile material was a problem during 1965, by 1968 the reprocessing plant in Trombay had started producing weapon-grade plutonium (Chengappa 2002: 111–12). A dedicated team of scientists under Rajagopal Chidambaram was earmarked to work on the 'equation of state' of plutonium necessary to determine its behaviour under impact of conventional explosives.[15] This was one of the most difficult scientific and technical problems in implosion devices. Also, in 1970, the DAE began the construction of a fast reactor called PURNIMA

[13] NMML (1967c).

[14] Abraham (1998) provides an excellent account on the role of the scientists in pushing India's nuclear explosion programme.

[15] This information is available in an interview of Dr R. Chidambaram, conducted by C.V. Sundaram, in September 1996. The interview is available in Sundaram et al. (1998).

(Sundaram et al. 1998). PURNIMA was vital for India's nuclear explosion programme as without it Indian scientists could not study the chain reactions produced by irradiation of plutonium. PURNIMA could be completed only in 1972 and it was only then that all technical necessities for a nuclear explosive device could be achieved by India's nuclear scientists.

However, even in 1968, when BARC seriously started preparing for nuclear explosive capability, no decision had yet been taken to actually conduct a nuclear explosion. The endeavour must be understood in light of the decision taken initially by Prime Minister Shastri, and later confirmed by Indira Gandhi, to fully achieve a 'nuclear option'. This did not translate into a decision to conduct a nuclear test. In fact, both in public and in private, the Indira Gandhi government adhered to a line that India would not produce nuclear weapons. India's nuclear energy programme was heavily dependent upon foreign assistance. In 1969, India completed its first power reactor in Tarapur with American assistance. In Rajasthan, with the help of Canada, India was building a series of power reactors. Any nuclear detonation would have jeopardized these extremely costly yet vital elements in India's nuclear energy development. Moreover, by the end of the 1960s, the threat of China had considerably reduced (Mukerjee 1968). For more than eight years since 1962, no major confrontation had occurred on the border.[16] India's conventional military capability had also increased after years of sustained defence modernization and development. If the principal lesson of the 1962 war with China was that Beijing will exercise 'military prudence' until and unless there is a 'certainty of military victory', India had achieved it by the late 1960s (Mehta 2010: 150). In fact, by 1968, the Indian military considered Chinese nuclear weapons a threat only in the 'long term' (Ministry of Defence [MoD] 1968: 1). Moreover, China itself was considerably weakened due to the Cultural Revolution, and also because of its conflict with the Soviet Union. The Indian decision–makers also remained convinced that a single nuclear detonation would not serve the purpose of India's security; a nuclear deterrent would require a vast wherewithal, including a large nuclear

---

[16] Except a skirmish on the border at Nathu La in September 1967 where Indian forces repulsed Chinese probes, with the People's Liberation Army suffering heavy casualties. The lesson of the Nathu la crisis, as the Indian Charge D'Affairs in Beijing reported to the Foreign office, was encouraging for New Delhi: 'The firmness shown by the Indian side in the skirmishes which occurred in the Sino–Sikkim border in September proved that those in charge of the cultural revolution had not lost all sense of proportion and respected force….China [now] shows considerable respect for India's growing military strength (NAI 1968)'.

arsenal and multiple delivery systems. As a top–secret note prepared by Sarabhai and Haksar for Indira Gandhi in April 1970 stated: ensuring India's security through a nuclear deterrent requires a 'total defensive system' involving a 'total commitment of national resources of a most stupendous magnitude'.[17] The Indian economy was just not prepared to incur such massive expenditures. Finally, the unspoken assumption in India's strategic thinking was that until and unless both the Soviet Union and the US had adversarial relations with China and continue to enjoy a favourable deterrence equation over her, the 'uncertainty' of their response in case of a Chinese nuclear threat against India would be an 'adequate deterrence' (Doctor 1971: 352).

By the early 1970s, India was moving closer to fully realizing the second leg of its response to the threat posed by the Chinese nuclear test in 1964: one of developing a 'nuclear option'. Slowly but surely, the scientific enclave was on the verge of mastering all relevant technological expertise required for a nuclear explosive capability. However, the decision to exercise the 'nuclear option' was not in the offing; at least not yet. The decision to conduct a nuclear test was also put off by events in India's neighbourhood. The war of Bangladesh's liberation overshadowed all other

[17] NAI (1970).

strategic considerations. After the war, India emerged as South Asia's military hegemon. It is also after the war that Prime Minister Gandhi finally gave her go-ahead for India's first nuclear test.

## The Peaceful Nuclear Explosion of 1974

If by the early 1970s, India was inching closer to realizing its nuclear option, the war in December 1971 provided it a security rationale to exercise it. The Bangladesh Liberation War significantly altered India's strategic environment. Following the 1971 war, India emerged as a regional power in South Asia. Yet, this war had left an impression of vulnerability on India's strategic mindset. For one, if after 1964 India had relied upon the US political guarantees, though implicit, to thwart the threat from China, under President Nixon the erstwhile arch-enemies had colluded against India during the 1971 war. In fact, the US had categorically told the Indian government in early 1970s that it would not be able to intervene to help India in case of a confrontation with China. This was part of the US policy to wean China away from the Soviet Union but left India deeply sceptical of President Nixon's foreign policy. If the US–China collusion was a dramatic shift in India's threat environment, Indo-US relations too reached their nadir when President Nixon dispatched the Seventh Fleet to the Indian Ocean in order to militarily coerce India from pursuing the war in East Pakistan (Nayyar 1977). In the post-war period, therefore, the triumphant assertion of India's regional supremacy sat uncomfortably with a sense of vulnerability vis-à-vis extra-regional interventions. It was this feeling of vulnerability at the hands of the US and the changing strategic scenario with the US–China collusion that may have led Indira Gandhi to finally give the go-ahead for India's first nuclear test in May 1972. At the least, a test would have signalled India's determination to be the leading South Asian power and increased its international status. It would have also increased Indira's grip on India.

The design of the nuclear device was ready by late 1971 (Perkovich 1999: 156). With the commissioning of the PURNIMA reactor in 1972, the Indian scientists now had the wherewithal to verify the design and enhance it further. Simultaneously, with the decision having been taken at the level of the prime minister, frantic activity was carried out by BARC scientists to prepare for the test. Pokhran, a sleepy town in the Thar Desert, was chosen as the test site. By September 1973, physical preparations at the test site had begun which included digging of shafts for placement of the device. Yet, the exact timing of the test was still being contested at the PMO. For one, Indira Gandhi's principal secretary, Prithvi Narain Dhar, an economist by profession, argued against explosion due to likely negative consequences for the Indian economy. Another confidant of the prime minister—P.N. Haksar—took the position that the test should be postponed until the next elections in order to reap political dividends. However, in 1974, scientists won the debate and the PNE went ahead in May. Yet, the exact date of the explosion provided Gandhi political mileage. This was precisely the time when the government was facing huge labour protests. Whether the explosion was motivated by reasons of security or by compulsions of domestic politics continues to generate extensive debate among India's nuclear historians.

Publicly, the government explained India's nuclear test as a PNE. It steadfastly refused to label the 1974 test as a nuclear weapon. The scientific community went to great lengths to explain how the test would help India's economic development (Ramanna 1978). The PNEs were considered to be useful for major infrastructural projects such as building dams, excavations for minerals, and even drilling of oil wells. The post-explosion rationalization of India's PNE therefore defied the main logic on which the 'nuclear option' strategy was accepted and supported by previous Indian decision-makers: as a hedge against insecurity created by China's nuclear weapons programme.

Whatever may be the real motivations for India's PNE, the device used in the May 1974 explosion was not weaponized. As Gaurav Kampani (2014b) has elaborated elsewhere, there is a fundamental difference between exploding a nuclear device and the process of weaponization. India's PNE involved a nuclear device which, by definition, can only be considered as an 'apparatus that presents proof of scientific principle that explosion will occur'. By contrast, weaponization entails a 'rugged and miniaturized version of the device'. It also involves a 'process of integrating the nuclear weapon with delivery systems' and its 'operationalization' with the development of soft institutional and organizational routines' for deployment and use. The 1974 explosion fit none of these criteria.

Revelations from archival research indicate, as noted, that the 1974 PNE was not a weapon and that India's leaders did not follow it up with concerted efforts to weaponize India's nuclear capability or develop the necessary delivery systems. This is most evident in a top-secret communication between India's MoD and Ministry of Finance (MoF) in January 1975.[18] The MoF claimed that the 1974 test provided India a rudimentary nuclear capability and therefore argued that the MoD

[18] For a detailed explanation, see Joshi (2015).

should not seek more funds for defense purposes.[19] The MoD was simply aghast. It complained about the MoF's misguided views to India's highest decision-making body, the Cabinet Committee on Political Affairs, which functioned directly under the prime minister. India's defence mandarins argued that it was 'unfortunate' for the MoF to have even made a 'mention of the nuclear blast' as it 'plays no part in our defence preparedness which is based entirely on conventional weapons'.[20]

For the MoD, the problem of associating the 1974 test with a nuclear capability was extremely disconcerting for three reasons. First, India's nuclear policy was focused exclusively on energy for peaceful purposes. Second, any use of nuclear weapons would bring India international opprobrium. Last, and the most important, was the fact that India had no demonstrated capability in nuclear warheads or delivery vehicles. The MoD noted: '[W]e cannot take into account the impact of our nuclear explosion on the [conventional] threat from Pakistan in the absence of [a] tactical nuclear weapon and a delivery system for it.' One can look at it as a bureaucratic struggle for scarce resources but the weight of the evidence suggests that the MoD was more accurate in its assessments of India's nuclear

[19] NMML (1974a).
[20] NMML (1975).

capability. New research indicates that no delivery system was being pursued by India in the 1970s which could have delivered a nuclear payload (Joshi 2015). Through the mid-1970s, India lacked a strategic air arm capable of delivering nuclear weapons. The Indian Air Force (IAF) neither envisaged such a strategic role for itself nor was it preparing to develop one. This must be seen in light of the fact that even the MoD was unsure of the strategic implications of the PNE. Some work on developing missiles was being carried out by the Defence Research and Development Organisation (DRDO), but none of the missiles envisaged at this stage were ballistic missiles capable of carrying a nuclear payload. At the end, the PNE could have provided India only an 'imagined arsenal'.

★★★

China's nuclear test of October 1964 forced India to seriously debate the threat posed by nuclear weapons to its national security. India's response to the Chinese nuclear threat, however, did not translate into full-blown indigenous nuclear weapons programme. India first relied upon a diplomatic strategy where it sought nuclear security guarantees from established nuclear powers. This strategy was dovetailed with India's championing of the NPT, at least during the early years of treaty. However, India's diplomatic crusade against Chinese nuclear threat failed as the NPT could not satisfy India's security concerns. India, therefore, opted to keep its nuclear option open. The 'nuclear option' entailed creating the necessary infrastructure to undertake a nuclear explosion in case India required to showcase its nuclear capability. In 1974, the nuclear option was finally exercised by New Delhi. However, India labelled it as a PNE. Its nuclear capability remained non-weaponized.

# 3

# The Failed Policy of
# Nuclear Refrain

India's PNE in May 1974 had proved its technical and
scientific capability. Though it had finally broken the
monopoly over explosive nuclear technology which
the NPT had enshrined upon the five nuclear weapons
powers, India did not initiate a full-fledged nuclear
weapons programme for another decade. In the history
of nuclear weapons worldwide, this was unprecedented.
In the aftermath of the PNE, India became the primary
target of the non-proliferation regime, spearheaded by
the advanced nuclear technology states in the West. As
a consequence, its nuclear energy programme suffered
greatly. This was also the period when Pakistan initiated
an intense quest for nuclear weapons. A Pakistani

Some sections of this chapter are based on Joshi (2018).

nuclear programme was the worst nightmare for Indian decision-makers because, unlike China, Islamabad suffered not only from a chronic insecurity vis-à-vis India but also pursued a revisionist design in South Asia. It had by now initiated and lost three major conventional wars. Nuclear weapons provided a major boost to Pakistan's revanchist agenda in Kashmir and beyond. It was also the most serious challenge to India's hegemony in South Asia. Even though New Delhi had proved its technical capability in 1974 and had conscientiously restrained from developing nuclear weapons, progress made by Pakistan in its quest for a nuclear capability forced India's hands. Also, more so because the non-proliferation regime and its votaries in the West not only ignored Pakistan's nuclear march but also provided her with enormous military assistance to fight the Soviets in Afghanistan. All these developments portended a grim future for India's national security. Early 1980s, therefore, saw a renewal of India's nuclear weapons programme. Thus began the nuclearization of the subcontinent.

This chapter traces these trends and their impact on India's nuclear policy between 1974 and 1984. It focuses on three major developments which confronted India's nuclear policy during this period. First of these developments was the strengthening of global non-proliferation regime. India's PNE provided the advanced nuclear technology countries a rallying point to tighten the gaps in the NPT through selective use of sanctions and safeguards. Their monopoly on advanced nuclear technology provided them with a stick to hammer the non-nuclear states to submit to their non-proliferation diktats. India was its first victim. This is the focus of the first section. Second was the initiation and progress of the Pakistani nuclear programme. If the defeat in the 1971 war had provided Islamabad a motivation to pursue nuclear weapons, India's PNE provided it a public alibi for its nuclear programme and further intensified its quest for nuclear weapons. China's assistance to Pakistan only confirmed India's worst fears: of a double encirclement by its two hostile neighbours. This is dealt with in the second section. Third, as the security scenario in the subcontinent underwent major transformation with the Soviet invasion of Afghanistan and the US support to Pakistan in containing communism, India earnestly began thinking of how to counter the Pakistani nuclear threat. The policy choices ranged from denuclearization of the subcontinent to pre-empting Pakistani nuclear programme through the use of military force. In the end, India resorted to develop its own nuclear deterrent. It veered towards a second series of nuclear tests and initiated a weaponization programme, including development of nuclear delivery systems. This is the focus of the third section.

# Non-proliferation Regime's Favourite Target

India's PNE of May 1974 evoked both hope and despair. The non-aligned world welcomed India's scientific and technological achievement—a Third World country had finally broken into the club of nuclear elites. If the Chinese nuclear test had dented India's prestige, the PNE had helped restore it. This was evident in the fact that a dozen countries as diverse as Saudi Arabia, Brazil, Argentina, Peru, Turkey, Libya, and South Korea sought India's cooperation in peaceful uses of nuclear energy after the PNE. India's insistence on using nuclear technology for peaceful purposes even after the test allowed her to sustain her moral leadership on nuclear disarmament. Unlike other nuclear powers, India could claim a moral high ground by eschewing the option of nuclear weapons.

The advanced nuclear technology states, however, saw in India's actions a challenge to the NPT and the principle of non-proliferation (Halsted 1974). Under the NPT, there was no distinction between a nuclear weapons explosion and peaceful nuclear test; but India did not accept this formulation for a number of reasons (Subramanian 1975). First, rather than capability, what mattered most in proliferation was a state's intention. Second, if India's intentions were peaceful, it was vital for a developing country like India to exploit all avenues of peaceful uses of nuclear energy for its long-term economic progress. Third, pursuing the PNE was a challenge to the inequitable world order established by the NPT where only a select few had the right to nuclear explosive technology. Such nuclear apartheid was unacceptable to India.

Notwithstanding New Delhi's reasoning, India's example was a bad precedent for non-proliferation and the NPT, especially as India's nuclear programme was assisted by technology transfers from the West. In the 1950s and 1960s, the UK, the US, and Canada had helped India's nuclear programme. The PNE, at least for the West, proved that India had misused the technology received for peaceful uses of nuclear energy. For non-proliferation, such pilferage had to be stopped. If the misuse of technology transfers was one major concern, another was the export of nuclear technology. Proponents of the NPT also wondered whether India would export explosive nuclear technology to other countries, defeating the purpose of non-proliferation. Therefore, the PNE was seen as a diplomatic challenge to the NPT. The first NPT review conference was due in 1975 and any more tests by New Delhi would have seriously jeopardized the treaty's future. These concerns guided the behaviour of advanced technology states towards India after the PNE.

Just a week after the PNE, Canada suspended all nuclear cooperation with India. Ottawa demanded that New Delhi place its nuclear facilities under the IAEA safeguards, accede to the NPT, and publicly declare that it will not transfer nuclear technology to other states (Keeley 1980). The US, however, was much more circumspect. Its principle interest, as Henry Kissinger explained to the Indian ambassador in Washington, DC, was in export controls.[1] India's response was guided by two contradictory factors. On the one hand were its long-held principles, such as the right to peaceful uses of nuclear energy (including PNEs) and the right to nuclear technology cooperation for peaceful purposes, to which India had to remain committed. On the other hand were practical considerations of its nuclear energy programme which was dependent upon foreign technology transfers. In this contest between principles and necessities, India choose the later. With Canada, India entered protracted negotiations during the course of which New Delhi accepted enhanced safeguards on nuclear technology and material provided by Canada (Kapur 1978). Both Rajasthan Atomic Power Station (RAPS) reactors were put under life-long safeguards monitored by the IAEA. In private, New Delhi also promised not to transfer nuclear explosive technology

[1] NMML (1974b).

to other countries and observe stringent export controls on nuclear technology. Indira Gandhi also promised that India will not conduct further PNEs till the time the data from the first test was fully studied, but this was just an excuse. After the PNE, Gandhi had made up her mind against further nuclear testing.[2]

However, India's private cooperation brought her no major dividends. The objective of the non-proliferation lobby was not to seek India's compliance but to make the non-proliferation regime foolproof. In April 1975, a group of 20 advanced nuclear states met in London to formulate stricter rules for nuclear technology transfers. Called the London Group, which later resulted in the NSG, these states issued a private trigger list of nuclear technology and materials. It was also decided to observe tougher supply conditions for their exports. India was its primary target. By May 1976, India's negotiations with Canada also failed. Ottawa terminated all its peaceful nuclear cooperation agreements with India.

The implications were severe for India's nuclear energy programme. Operation of Canadian-supplied RAPS-1 reactor suffered and RAPS-2, which was under construction with Canadian assistance, was

---

[2] Interview with a senior retired Indian diplomat, New Delhi, September 2017.

severely delayed. Most importantly, India's entire indigenous reactor programme, which was largely based on the Canadian CANDU-type reactor technology, was threatened. The problem was India's material capability. Even after a massive R&D programme pertaining to nuclear energy, India had no industrial base to produce nuclear components. Its heavy water programme, even after years of R&D investment, was yet to deliver. In fissile material, India was yet to master uranium enrichment needed for Tarapur-type light water reactors. Restrictions imposed after the PNE, therefore, crippled India's nuclear energy programme. It also led to a massive effort towards creating an indigenous industrial base for producing nuclear components, largely by reverse engineering the material and technology procured from foreign vendors in the past.

This friction with the emerging non-proliferation regime continued even after the ouster of Prime Minister Gandhi and the coming of the Morarji Desai government in 1977 (Noorani 1978). Desai, while in opposition, had criticized Indira Gandhi's nuclear policy and was adamantly against a nuclear weapons programme, even an ambivalent one.[3] He was also not

[3] For Desai's nuclear policy, see Gupta (1978) and Kapur (1978).

convinced of Gandhi's adversarial approach towards the West and her rather sympathetic attitude towards the Soviet Union.[4] In fact, Desai had promised to make India genuinely non-aligned, by which he meant retracting India's special relationship with the Soviet Union and focusing upon improving India's relations with the Western countries. Soon after becoming the prime minister, Desai declared that India will not conduct any more PNEs, a commitment Indira Gandhi was not ready to give, at least in public. Desai's policies also endeared him to President Carter, who not only wanted to improve India–US relations but also extract India's support for his policy on non-proliferation. During the first year of Desai and Carter's term in power, India and the US initiated a serious dialogue on reconciling their nuclear differences, including full-scope safeguards on India's nuclear facilities.

This brief bonhomie between India and the US, however, fell prey to domestic politics in the two countries. In mid-1977, the US Congress had enacted the Nuclear Non-proliferation Act (NNPA) which directed the administration to stop supplies of nuclear material and technology without full-scope safeguards (Chari 1978). This directly affected the operations

[4] For a discussion on Desai's security policy, see Thomas (1980).

of the Tarapur nuclear power plant because under the 1963 agreement, the US had promised to supply highly enriched uranium fuel for the lifetime of the reactors. For India, the implementation of NNPA was a direct violation of a bilateral treaty which had been signed much before. It was also seen as a blackmail: the US was using its supplier status to coerce India into accepting full-scope safeguards (Kapur 1979). Even when Desai publicly articulated an aversion to nuclear weapons and tried to reconcile differences with the US, it was politically difficult for him to relent to the new conditions as the political opposition accused him of relinquishing India's strategic autonomy. By 1978, delayed shipments of nuclear fuel from the US became a constant rallying point for the opposition. As the date of implementation of the NNPA neared in 1979, the brief bonhomie in India–US relations also fell apart. Desai's position in the Janata coalition had by then weakened considerably, further denting his credentials to take any policy decisions on India's nuclear future. If Desai's hands were tied domestically, one external factor with far-reaching consequences for India's nuclear policy was fast developing in her immediate neighbourhood. By 1979, Indian intelligence confirmed Pakistan's successful mastering of uranium enrichment technology. The spectre of a Pakistani nuclear programme was the final nail in the coffin of

Desai's brief agreement with the non-proliferation regime.

## The Rise of a Nuclear Pakistan

In May 1965, Zulfikar Ali Bhutto, in an interview to the *Manchester Guardian*, had argued that if India got a nuclear bomb, 'we [Pakistan] will eat grass, even go hungry but we will get one of our own [nuclear bomb]' (cited in Khan 2014: 59). The context was the growing clamour in New Delhi to go nuclear after China's October 1964 nuclear test. As was argued in the previous chapter, for ten long years until May 1974, India refused to satisfy the necessary condition for Pakistan's nuclearization laid down by Bhutto in 1965. However, as Adrian Levy and Catherine Scott-Clark (2007: 18–19) have argued, Pakistan had found a reason to go nuclear after 1971: the war in Bangladesh had left Pakistan bifurcated. The beginning of the Pakistani nuclear weapons programme in the early 1970s was a result of Islamabad's conventional insecurity. Where it fell short in conventional power, nuclear weapons were deemed necessary to fill the void: the existence of nuclear weapons could have allowed Pakistan to avoid another 1971-type military defeat. It was for this reason that Prime Minister Bhutto decided to give the go-ahead for a Pakistani nuclear programme in

1972. New archival research indicates that India was oblivious of Pakistan's nuclear efforts before the 1974 PNE (Joshi 2015). In India's defence planning, the Pakistani threat remained purely conventional. To allay Islamabad's concern after the PNE, Prime Minister Gandhi had assured Prime Minister Bhutto that New Delhi will remain committed to its 'traditional policy of developing nuclear energy for peaceful purposes'.[5] However, India's 1974 PNE provided Pakistan a rationale to legitimize its nuclear aspirations. As Prime Minister Bhutto argued in December 1974, 'If Pakistan is not able to acquire weapons (conventional), which can act as a deterrent, it must forgo spending on conventional weapons and make a big jump forward concentrating all its energies on acquiring the nuclear capability' (Jayagopal 1978: 190).

Pakistan initially chose the more difficult route of developing plutonium-based weapons. For this, Islamabad had to develop a reprocessing capability. Given the complexity of this technology and Pakistan's nascent nuclear programme, it was only possible through external assistance. Pakistan signed a deal

[5] US Department of State, Bureau of Near Eastern and South Asian Affairs, 'Indira Gandhi's Letter to Bhutto', 29 May 1974, available at https://wikileaks.org/plusd/cables/1974NEWDE07109_b.html.

with a French firm to construct a reprocessing plant at Chasma in October 1974. Bhutto explained these developments as a part of Pakistan's nuclear energy programme. During his visit to the US in early 1975, the Pakistani prime minister promised Americans that it will remain committed to a peaceful nuclear programme. However, Bhutto also used the threat of India's nuclear programme as an excuse to extract further military assistance from the US, which Washington agreed to. Pakistani efforts to develop plutonium fuel cycle came under the scrutiny of the emerging non-proliferation regime, which specifically targeted transfer of enrichment and reprocessing technologies to non-nuclear states. As the US pressure on France increased after the formation of the London Group in April 1975, the French demanded more stringent safeguards. Pakistan had no choice but to accept the French demands. However, in 1975, a new element radically changed Pakistan's nuclear fortunes: A.Q. Khan, who was arguing that Pakistan should undertake the uranium enrichment route for a nuclear weapons programme. In 1976, A.Q. Khan returned to Pakistan to work on its nuclear weapons programme with stolen designs of the centrifuge-based uranium enrichment technology from the European nuclear consortium, URENCO. Khan thereupon took charge of the Pakistani weapons programme and established a worldwide proliferation network to source material for Pakistan's uranium enrichment plant at Kahuta.

After the French reprocessing deal, India's intelligence services began scrutinizing the progress of Pakistan's nuclear programme more closely. The first assessment of Pakistani nuclear progra mme was made by India's Joint Intelligence Committee (JIC) in March 1975.[6] This intelligence estimate made little of Pakistani nuclear capability and concluded that it will take Islamabad decades to develop nuclear weapons. By 1976, the Indian intelligence was convinced that Pakistan's quest for a reprocessing plant was being severely undermined by the emerging non-proliferation regime.[7] Ironically, the non-proliferation regime which was triggered by India's PNE was now targeting its main adversary. Even when India was fighting its own battles with non-proliferation cartels like the NSG, it also saw in them a means for restricting Pakistani nuclear programme. However, these intelligence estimates were flawed to the extent that they exclusively focused on Pakistan's quest for plutonium reprocessing technology.[8] Till at least early 1978, India remained unaware of the

[6] NAI (1976).
[7] NAI (1976).
[8] NAI (1977a).

A.Q. Khan network.[9] India was, however, worried about China's assistance to Pakistan, and rightly so. Not only was China a close ally of Islamabad but was also unencumbered by the NPT; it had no stake in restricting nuclear proliferation. Moreover, a Pakistani nuclear programme could have seriously undermined India's national security and challenged its pre-eminence in South Asia. As the JIC report stated, 'If Pakistan at all succeeds in exploding a nuclear device within the next five years or so, it will probably be with the help of China which is the only nuclear power that is not the member of the IAEA and is known to oppose nuclear hegemony.' These concerns were borne out in June 1976 when Bhutto was able to finally get Chinese assistance in the design of a nuclear device. As Bhutto later described in his autobiography, securing Chinese assistance for Pakistani nuclear programme was the 'greatest achievement' of his entire political career (Bhutto 1979: 223).

[9] For when Indian intelligence came to know about Kahuta enrichment plant, see Kasturi (1995). One of the reasons for this intelligence blackout was that the new Morarji Desai government had severely curtailed the finances and functioning of India's external intelligence agency, Research and Analysis Wing (RAW). See Raina (1981).

New Delhi had failed to accurately assess the progress of the Pakistani nuclear programme in the mid-1970s. The coming of the Carter administration in Washington, DC, and its non-proliferation policy, which specifically targeted plutonium reprocessing further strengthened a perception that nuclear proliferation would henceforth become exceedingly difficult. By mid-1977, the Pakistani–French reprocessing deal appeared to be dithering under the collective impact of the US non-proliferation policy and domestic political changes in both France and Pakistan.[10] Other Western countries like Canada also tied their nuclear cooperation with Pakistan on nuclear safeguards and revising its reprocessing plans.[11] Between 1976 and 1978, India was willing to amend its relations with Pakistan with a hope of addressing the latter's insecurity. In March 1976, the Indian government, with an aim to achieve a 'comprehensive normalization' of its relationship with Pakistan, approved restoration of full diplomatic relations severed after the 1971 Bangladesh War (Mehta 2010: 163). When in early 1977 Indira Gandhi lost the elections and Morarji Desai became prime minister, the process of reconciliation with Pakistan continued. In fact, Desai had even offered Pakistan assistance in the

[10] NAI (1977b).
[11] NAI (1977c).

development of peaceful uses of nuclear energy. Such diplomatic overtures notwithstanding, it was also the period when Pakistani efforts in uranium enrichment and nuclear weapons design made the most progress.

In the post-PNE period, therefore, India had reposed its faith in the emerging non-proliferation regime to arrest the Pakistani drive towards a nuclear weapons programme. In public, India maintained its opposition to it; yet, for its own security imperatives, India was banking on the 'lesser evil' of a discriminatory non-proliferation regime (Mehta 2008: 425). And it was the weakness of the same that helped Pakistan make steady progress. Even when the mandarins in India's intelligence community had clearly identified the China angle, they fell short in fully comprehending the A.Q. Khan factor. By 1979, however, India was in for a major surprise. Though Indian intelligence got wind of the A.Q. Khan network by 1978, it was only in April 1979 that the JIC revised its earlier views on Pakistani nuclear capabilities and argued that Pakistan would be able to explode a nuclear device in a year or two (Subrahmanyam 1998a: 36). This was the most dramatic development in India's security environment after the 1964 Chinese nuclear weapons test. Indian decision-makers were, in fact, completely taken by surprise.

Within five years of the PNE, Pakistan was on the verge of a nuclear weapons capability. India, however, had deliberately eschewed developing a full-fledged nuclear weapons programme even after proving its technological capability. On the one hand, New Delhi tried to allay Pakistan's insecurity through diplomatic means. Both Indira Gandhi and Morarji Desai believed that negotiations would help stabilize India–Pakistan relations. India's moral aversion to nuclear weapons also played a role, though to a lesser degree in the case of Indira Gandhi. On the other hand, India also believed that the emerging non-proliferation regime would arrest Pakistani nuclear programme. As India's Foreign Minister Vajpayee complained to the US Secretary of State Cyrus Vance in April 1979, 'how it was that inspite to laws and safeguards, Pakistan had managed to move ahead in acquiring a nuclear weapon capability'. The answer was simple: the safeguards regime was far from foolproof. India had placed its bets on the wrong horse.

## Whither Nuclear Restraint

Indira's return to power in December 1979 coincided with some of the most portentous developments in India's security environment, creating pressures on its nuclear policy (Gupta 1983). Pakistan's march towards a nuclear weapons capability was now only a matter of time; its intentions and trajectory were set. In

December 1979, Soviet Union invaded Afghanistan to support the pro-communist regime of Babrak Karmal. India had always opposed great power conflicts the world over and now, for the first time, a great power had intervened in India's immediate neighbourhood.[12] Soviet invasion of Afghanistan threatened America's interests in the Persian Gulf. It also provided the US an opportunity to repay the favours it had received during the Vietnam War by inciting an armed insurgency in Afghanistan. Pakistan's readiness to be the face of the US resistance to Soviet occupation made South Asia the hotbed of Soviet–American animosity. In return, Pakistan extracted military and economic aid from Washington. This event also created hardships for India's policy of non-alignment. Soviet Union was India's principal military and economic partner. The expectation of India's diplomatic, if not material, support was therefore obvious. Yet, India's silence on Soviet intervention would have been perceived as duplicitous by the West.

Indira's approach was defined by India's national interests. To America's dismay, India declined to condemn Soviet occupation of Afghanistan at the UN. India's dependence on Soviet military hardware

---

[12] For an Indian assessment of the Soviet invasion of Afghanistan, see Dixit (2002).

was a major consideration. It acquired further importance as Pakistan started receiving military and economic aid from the US to fight Soviet forces in Afghanistan during Carter administration's last year in power. This threatened the conventional military balance in South Asia, which had traditionally been to India's advantage. With the coming of President Reagan in 1981, the US policy in Afghanistan pivoted completely around Islamabad. Reagan's rabid anti-communism threw the semblance of balance which the Carter administration had tried to achieve in balancing India's concerns vis-à-vis its support for Islamabad in the fight against Soviets. In April 1981, the US announced a \$3.2 billion military and economic aid for Pakistan (Kux 2001: 258). This included top-of-the-line F-16 fighter jets, which were capable of delivering nuclear weapons. As a secret note prepared by the MEA observed, this arms build-up was a 'serious aggravation in its [India's] security environment'.[13] This blatant disregard for Indian security concerns underlined a change in America's non-proliferation policy. The Reagan administration had reached a tacit bargain with Pakistan on the latter's nuclear weapons policy: sans nuclear test, America would turn a blind eye to Pakistan's nuclear

[13] NAI (1982).

weapons programme. These developments occurred at a time when India considered a Pakistani nuclear test imminent. A secret assessment of Pakistani nuclear capability made by Indian intelligence in early 1981 estimated that Pakistan had amassed enough uranium to undertake a nuclear test within a year and it was also possible that it had mastered the technology of nuclear triggers.[14] This led to a conclusion that all 'main technical elements of conducting a nuclear test are already available in Pakistan'. Equally disconcerting for India was Chinese collusion: as the memo argued, 'the Chinese connection to Pakistani programme was less in the realm of speculation than in the realm of reality'. It also stated that Pakistan may conduct its first nuclear explosion at Lop Nor as 'it would appear to be just one more nuclear test by PRC' and help her avoid 'the range instrumentation problems'. These developments engendered a serious debate within to resuscitate India's dormant nuclear explosive programme. For the first time, the Indian military also openly started debating the consequences of South Asia's nuclearization and its impact on India's conventional military deterrent.

India's first response was of a pre-emptive military strike against Pakistan's uranium enrichment facility at

[14] NAI (1981). All quotes in the para are from this source.

Kahuta. In 1981, the Israeli Air Force had successfully destroyed an Iraqi nuclear reactor at Osirak. According to a senior scientist who was then studying Pakistani nuclear programme for Indian intelligence, Prime Minister Indira Gandhi did seriously consider this option in late 1982.[15] The acquisition of Jaguars, a deep penetration strike aircraft, from Britain in the late 1970s provided the IAF the requisite firepower to target Pakistani nuclear facilities. The problem, however, was Pakistani reaction. Islamabad could declare a full-blown war upon India and the international community would have blamed New Delhi for its initiation. Islamabad could also have targeted India's nuclear facilities in a tit-for-tat move, especially the BARC at Trombay. Given India's larger nuclear infrastructure, the radioactive spillage could have been catastrophic. These concerns precluded the pre-emptive strike to emerge as a policy choice. At least in the minds of Indian decision-makers, a rudimentary nuclear deterrence was now a factor in India–Pakistan relationship.

Instead, India veered towards a nuclear test (Central Intelligence Agency [CIA] 1981: 1). In her second innings as the prime minister, Indira Gandhi realized that restraint exercised after the 1974 PNE was indeed

[15] Interview with Dr K. Santhanam, New Delhi, 16 December 2015.

a mistake.[16] Soon after assuming power, she tried to undo the anti-nuclear policies of Morarji Desai. She rescinded the promise made by her predecessor that India will not conduct any further nuclear tests. If Desai had taken India's nuclear scientists to task, Indira restored the autonomy and power of BARC. In 1981, Raja Ramanna was sent to BARC as director, with a purported intention to revive India's nuclear explosion programme. She had also asked the nuclear scientists to remain prepared for a second nuclear test, which included maintaining the nuclear shafts at the Pokhran testing range in Rajasthan. Though the exact timing is contested, sometime in 1983, the Indian prime minister gave a serious thought to a second nuclear test. The then defence minister, R. Venkataraman, later acknowledged that new shafts were dug at Pokhran and all preparations were completed (Venkataraman 1998). However, Gandhi demurred at the last moment. It is still unclear why she backtracked, but the fear of economic sanctions seems to have played a significant role in her calculus. Also, a nuclear test would have given Pakistan an opportunity to undertake its own nuclear explosion, with India receiving only international opprobrium for forcing Pakistan's hands.

[16] Interview with a senior Indian diplomat, New Delhi, September 2017.

India, therefore, prepared to weaponize its nuclear option, which entailed preparations for eventual delivery of nuclear bombs. Though India had exploded a nuclear device in 1974, as has been described earlier, it was not usable as a weapon. That requires miniaturization and further design improvements so as to be able to be delivered through aircraft or missiles. In 1974 and afterwards, no delivery systems were available, nor was New Delhi developing any. The pursuit of delivery vehicles began in earnest in Indira's second term. In 1983, India launched the Integrated Guided Missile Development Programme (IGMDP), under which ballistic missiles such as Prithvi and Agni were to be developed (Shukla 2013). These, in future, would become the preferred system for nuclear weapons delivery. However, as the IGMDP's fruition was bound to take substantial time, India's DRDO started experimenting with available aircraft in the inventory of the air force as delivery vehicles. The DRDO first tested its nuclear weapon designs on the Jaguar aircraft in 1984 but found their ground clearance too shallow to carry a nuclear payload in its underbelly (Kampani 2014b). India's quest for its first nuclear delivery vehicle stopped with the French Mirage aircraft, which were ordered in early 1980s but were yet to be commissioned. Notwithstanding these initial problems, the fact was that by the time Gandhi's prime ministership ended with her assassination in October 1984, India was veering towards weaponization of its nuclear capability. This process would be advanced by her son and political heir, Rajiv Gandhi.

★★★

The non-weaponization of the 1974 nuclear explosion proved that India was a reluctant nuclear power. India is the only country in the history of nuclear proliferation that did not immediately produce nuclear weapons after conducting a nuclear test. It, instead, insisted that the PNE was part of its quest to develop peaceful uses of nuclear energy. India's intentions did not bring her any great advantages. In the aftermath of the PNE, the country became the non-proliferation regime's principal target. Advanced nuclear states imposed technology restrictions, thereby crippling its nuclear energy programme. Nuclear suppliers like the US conditioned further supplies of fissile material on India's acceptance of more stringent safeguards regime. Even when India was the target of the emerging consensus on non-proliferation among the advanced nuclear states and technology cartels such as the NSG, it did comply with the precepts of non-proliferation while continuing its moral opposition to such an inequitable bargain. It eschewed transfer of nuclear explosive technology and observed strict export controls. Yet, New Delhi maintained a stoic resistance against giving up its nuclear option by not accepting full-scope safeguards. The problem for India was that non-proliferation was both a liability and also an asset: liability when it came to constraints on its nuclear option and an asset when it helped curb proliferation in its neighbourhood. In fact, as this chapter has argued, India did repose its faith in the non-proliferation regime to curb Pakistan's nuclear programme between 1974 and 1979. In the end, the weakness of the same allowed Pakistan to develop a nuclear weapons programme. The lesson New Delhi learnt was that the decision not to produce nuclear weapons after 1974 was indeed a mistake. This was compounded by the security environment in which India found itself in the early 1980s. Great power conflict finally arrived at its doorsteps with the Soviet invasion of Afghanistan. This created enormous problems for India's security as the US sought Pakistan's help in fighting Soviet communism, in the process doling out billions of dollars for Pakistan's military build-up. It also turned a blind eye to Pakistan's nuclear programme: geostrategic interests trumped America's non-proliferation policy. In such a situation, India resuscitated its dormant nuclear explosive programme, this time with the intention to produce nuclear weapons. It also initiated a long-term plan to develop a credible nuclear deterrent by initiating a ballistic missile programme. In the short term, it relied on improvising upon its fighter aircraft.

This decade of uncertainty in India's nuclear policy ended with only one certainty: of South Asia's nuclearization. Both India and Pakistan now veered towards an existential nuclear deterrence. In 1998, both these countries would openly declare themselves as nuclear weapon states. The process behind their overt nuclearization is covered in the next chapter.

# 4

# Pathway to a Nuclear Weapon State

By October 1984, when Prime Minister Indira Gandhi was assassinated, India had started moving gingerly towards a nuclear weapon capability. India's journey to becoming a full-fledged nuclear weapons power took another decade-and-a-half, however. Three factors shaped India's nuclear awakening. First, the threat of Pakistani nuclear weapons and Chinese collusion showed no signs of abating. In fact, under the shadow of nuclear weapons, Pakistan became overtly hostile, and started pursuing a proxy war in Kashmir and abetting terrorism across India's body politic. To counter the burgeoning nuclear threat in its neighbourhood, India had no choice but to have a nuclear deterrent of its own. Second, with the end of the Cold War and the disintegration of the Soviet Union in 1991, India's security environment undertook a dramatic turn. India lost its most trusted security partner during the Cold War, the Soviet Union, and the international system was now fully dominated by a single superpower, the US, which in the past had been unsympathetic, if not overtly hostile, to India's security concerns. In this unipolar world order, the US also became the prime patron of non-proliferation and exerted enormous pressure upon India to submit to the non-proliferation regime. Third, the shifts in domestic politics were equally important, especially the rise of BJP. Even when successive Indian prime ministers protected India's nuclear option, advanced R&D on nuclear weapons technology, and laid the foundations of a deliverable nuclear arsenal, it was only BJP which favoured an open declaration of India's nuclear capability. Where the Indian National Congress suffered from both a moral dilemma and a policy inertia set by its past leadership, BJP openly celebrated the power and prestige associated with nuclear weapons.

This chapter outlines India's nuclear policy between 1984 and 1998. It is divided into three sections. The first section discusses India's nuclear policy under Prime Minister Rajiv Gandhi. Rajiv Gandhi's prime ministership saw India making tangible progress in weaponizing its nuclear deterrent, a process which began under his predecessor, Indira Gandhi. The second section deliberates the impact of the end of the Cold War and of the strengthening of the non-proliferation regime on India's policy choices. The last section delves into the weapons tests of May 1998 and its immediate consequences.

## Rajiv Orders Full Weaponization

By the time Rajiv Gandhi became the prime minister in late 1984, Pakistan's nuclear programme had made steady progress. Indira Gandhi had begun the process of materially equipping India with a nuclear arsenal, but material capability is only one side of nuclear deterrence. Nuclear deterrence also requires the software of operationalization. The first challenge Rajiv faced as the new prime minister was to decide on the shape and size of India's nuclear deterrent. In the summer of 1985, the Minister of State for Defence Arun Singh formed an informal committee to deliberate on India's nuclear deterrent force.[1] This committee was headed by Lieutenant General K. Sunderji, the then vice chief of the army staff. The committee recommended a nuclear force of 60–130 warheads, primarily delivered

[1] Details on this committee can also be found in Kennedy (2011) and Perkovich (1999).

through the air vector.[2] The cost of the arsenal, in the committee's calculations, could have been anywhere between Rs 7,000 and 8,000 crores (70 and 80 billions). The committee argued that if given the go-ahead, this arsenal could be managed within three years by 1988–9. The report of this committee was the first-ever official articulation of India's nuclear thinking, which subsequently came to be known as 'minimum deterrence'. The report, however, did not result in any major follow-up by the Rajiv Gandhi government. Instead, the government explored diplomatic avenues to manage the increasing nuclear overtones in India–Pakistan equation. In December 1985, India and Pakistan signed their first major confidence-building measure (CBM) by agreeing not to attack each other's nuclear facilities. One reason for this was the fact that Pakistan was, by now, in possession of at least five to six nuclear weapon devices and had hence acquired an 'elementary nuclear capability' (Dixit 1995: 185).

That an existential nuclear deterrence was in effect in the subcontinent became fully evident in 1986–7 when India and Pakistan entered one of the most intense military crises South Asia had seen since the 1971 war. In the summer of 1986, the Indian Army

[2] Interview with a senior naval officer who was part of this committee, New Delhi, 16 October 2015.

conducted one of its biggest military exercise—Exercise Brasstacks—along the western border with Pakistan. Brasstacks was an attempt to signal India's increasing conventional strength to Islamabad. It, some argue, was also a ruse to initiate a major military confrontation with Pakistan so as to eliminate its nuclear capability. Whatever be the real motivations, India's military moves evoked major counter-mobilization from Islamabad. By December, the two militaries were staring at each other across the international border. Yet, the exercise was important because, for the first time, Islamabad not only openly accepted its nuclear capability but also issued a nuclear threat to India. As A.Q. Khan, father of the Pakistani nuclear programme, told an Indian journalist in January 1987, 'we shall use the bomb if our existence is threatened' (Nayar 1987). The crisis was defused by high-level diplomacy between Prime Minister Rajiv Gandhi and President Zia. The lessons of the crisis were evident. First, if Bhutto had initiated the Pakistani nuclear programme in 1972 to blunt India's conventional military capability, Brasstacks was its first manifestation. This would become the hallmark of Pakistani nuclear strategy henceforth. Second, conventional military activity between India and Pakistan now involved the risk of nuclear escalation. The problem for India, however, was that even when Pakistan was issuing nuclear threats, India's nuclear weapons programme had yet not acquired enough teeth as a deliverable nuclear arsenal was missing. As K. Subrahmanyam has argued in his memoirs, 'In the period 1987-1990, India was totally vulnerable to the Pakistani nuclear threat' (Subrahmanyam 1998a: 44).

Therefore, 1988 was one of the most crucial years in Indian nuclear decision-making. Two major decisions changed the pace of India's nuclear weapons programme. First, Prime Minister Gandhi made the last diplomatic argument for India's renunciation of nuclear weapons. In the third UN Special Session on Disarmament, Gandhi presented to the world leaders a comprehensive plan for nuclear disarmament (Gandhi 1988). Called the 20-year disarmament programme, it was by far the most important diplomatic effort made by India since the NPT to eliminate the threat of nuclear weapons. The action plan envisaged complete and verifiable nuclear disarmament in a time-bound framework. Rather than the abstract guarantee towards nuclear disarmament enshrined in NPT, the action plan, for the first time in the history of nuclear disarmament diplomacy, advocated a time-specific disarmament plan. India's disarmament diplomacy was, however, running against the tide of change in international politics. Though supported by the Soviet Union, it only received a lukewarm response from other nuclear powers, especially the US, which was much more interested in what India could do for its nuclear non–proliferation agenda than what Washington would do for India's disarmament goals (Subrahmanyam 1998a: 44). Even when India's pleas fell on deaf ears, it was in some sense a moral vindication for Prime Minister Rajiv Gandhi. He had made the last effort for the principle of disarmament which India had stood for since its independence; its security now required resorting to the ultima ratio of power in international politics, of nuclear weapons.

Second, suffering humiliation in the UN, Prime Minister Gandhi ordered the weaponization of India's nuclear capability. Since the beginning of his term as prime minister, Rajiv Gandhi had authorized the DRDO 'to start development of rugged, miniaturized, safer and more reliable components and subsystems for what might eventually be called a weapon system' (Kampani 2014b: 91). The idea was to maintain a nuclear capability at a 'minimum state of readiness'. After the humiliation at the UN, Rajiv allowed full development of an air-delivered nuclear arsenal. Fighter aircraft became the principal vector of India's nuclear delivery, principally because of two reasons. First, India's missile programme was far from ready to carry nuclear warheads. India's first ballistic missile, Prithvi, was test-fired for the first time only in February 1988, followed by the longer-range Agni missile in May 1989. However, both these systems were still in the

R&D stage. Second, even though Indian scientists had started miniaturization of nuclear warheads, they had not reached a stage where the weapons were compact enough to be carried on missiles. India had also started a nuclear submarine programme in the late 1970s, but it was geared towards producing nuclear-powered attack submarines (SSN) for the Indian Navy rather than nuclear ballistic missile submarines (SSBN). The only platform available to the scientists was therefore the aircraft. By the late 1980s, India acquired Mirage fighter aircraft from France and the DRDO started modifying these aircraft to carry a nuclear payload. The IAF also started training its pilots in the intricate manoeuvres called 'flip-toss' or 'bomb-toss' required for releasing the bombs from an aircraft's underbelly. It was indeed a primitive style of nuclear delivery; however, at least in the short term, this was the only option available.

Once the decision was made by Prime Minister Rajiv Gandhi, all successive governments continued the same trajectory. It had taken a lot of internal debate for India to choose a nuclear path; once decided, it also enjoyed a certain consensus. When in December 1989, V.P. Singh became prime minister, the process continued uninterrupted. Singh also faced growing nuclear threat from Pakistan when, in 1990, a major crisis erupted between India and Pakistan over the latter's support for terrorism in Jammu and Kashmir. The threat of nuclear use by Pakistan during the crisis may be over-exaggerated but given the nuclear overtones, Singh was forced to consider India's responses in case of a nuclear emergency. A new secret committee—called the Arun Singh Committee—now considered India's response to a Pakistani nuclear strike (Perkovich 1999: 313–14). Command and control of nuclear assets became important even as provisions were made for storage of delivery vehicles and nuclear warheads, and air force units were pre-designated to carry out nuclear strikes in case of deterrence breakdown. Forced to initiate weaponization of its nuclear option by a threatening nuclear adversary, India could not escape the supplementary steps required to establish a nuclear deterrent.

By the end of the 1980s, India had exploited all avenues of diplomacy to answer the threat posed by nuclear weapons, both in the region and in the world. In return, India received only failures and disappointments. Indian decision-makers were not starry-eyed idealists. Since Indira Gandhi's second term, Indian scientists were asked to make preparations for a deliverable nuclear arsenal. The moment of decision came in 1988 after Rajiv Gandhi's appeal for nuclear disarmament was ignored by the nuclear haves at the UN. India was now free of its moral obligation; and it was now ready to accept nuclear weapons as an important element of its national security strategy. The events in early 1990s further reinforced these trends in India's nuclear policy.

## Tightening Noose of Non-proliferation

The dissolution of Soviet Union in 1991 took away India's most important pillar of external support. Since the 1971 Treaty of Friendship, India had relied upon the Soviet Union for both military and diplomatic support. America's unipolar moment also became the 'age of nuclear enlightenment'. For the first time after the NPT process, a great power consensus emerged around making non-proliferation the benchmark of international security. The end of the Cold War also saw further strengthening of the technology denial regimes. By 1987, technologically advanced states had instituted the Missile Technology Control Regime (MTCR) which intended to restrict proliferation of ballistic missile technology.[3] Consequently, in 1992, the NSG issued a new set of guidelines (Part 2 of INFRIC/254), adding to the original set of guidelines, on export of dual-use items decided in 1978. This turn towards non-proliferation took a more aggressive turn with the coming of the Clinton administration at

[3] For a comprehensive history of MTCR, see Mistry (2003).

the White House in 1993. In a unipolar world where American power reigned supreme, India's nuclear weapons programme came under intense pressure as South Asia was considered the hotspot of proliferation. The goal for the Clinton administration was to 'cap, roll back and eliminate' India and Pakistan's nuclear weapons programme (Gordon 1994: 570–1).

This was also a period of immense political and economic uncertainty in India. The fiscal crisis of 1991, and the consequent liberalization of the Indian economy, made the country vulnerable to Western economic pressure because of loans from the World Bank and the International Monetary Fund (IMF). The assassination of Rajiv Gandhi in 1991 had made the political situation equally uncertain. It was under these circumstances that P.V. Narasimha Rao came to the centre stage of Indian politics and became prime minister. Rao had one of the most difficult jobs any Indian prime minister had ever faced. He had to balance India's economic interests with India's security, while his own position was compromised by both his coalition partners and infighting within the Congress. However, he managed India's multiple challenges deftly.

Under Prime Minister Rao, India's nuclear weapons programme continued apace and from being only an option till mid–1985, it could now better be understood as a posture of 'recessed deterrence': to have nuclear weapons without openly declaring possession of such a capability (Perkovich 1993). India would have acquiesced to an undeclared nuclear capability and an undisclosed nuclear arsenal only if its option to do so at a future date could have been maintained. However, the possibility was that India may not have this choice, as was identified by the Rao government just after the prime minister's meeting with the US President George H.W. Bush in January 1992. Pressure from the US was immense and Rao had agreed to initiate a nuclear dialogue with Washington, DC, in order to reconcile differences over India's nuclear weapons policy. However, Rao had no intention to submit to America's diktats on India's nuclear programme. His strategy was to bide time while India's nuclear weapons development reached a stage of maturity and the economy recovered to an extent that India could withstand the consequences of openly declaring itself a nuclear weapons power. Just after his parleys with President Bush in the summer of 1992, an informal committee under the PMO was constituted with a mandate to 'consider all aspects of India's nuclear and space policies and to recommend the negotiating stance to be adopted by the Government of India' (Dixit 1996: 373). One of the recommendations of this committee was to 'ensure the widest possible freedom of options to increase our nuclear, missile and space capabilities as time was running out because of the likelihood of punitive international regimes coming into force by 1996–97' (Dixit 1996: 373). This recommendation is interesting for two reasons. First, it suggests that negotiations on arms control treaties such as FMCT, CTBT, and the NPT were clearly a strong determinant in India's post-Cold War nuclear policy. Second, it also suggests a sense of desperation within the decision-makers regarding the acquisition of an effective nuclear and missile capability. Both these factors played a crucial role in India's nuclear policy.

The nuclear dialogue with the US made no major headway. By 1994, it was apparent that the goals of Washington, DC, and New Delhi diverged significantly. Furthermore, as more concrete proof of the collusion between China and Pakistan in missile and nuclear proliferation emerged in the public domain, India's insecurity vis-à-vis its two hostile neighbours only increased. It also underscored the failure of the non-proliferation regime and the US' inability to restrict the Sino-Pak collusion. When, under these circumstances, the nuclear weapons power pushed for an indefinite extension of the NPT in 1995, India made an impassioned appeal for extracting more concrete disarmament guarantees from the nuclear weapon states. However, in May 1995, more than 170 NNWS agreed to the permanent extension of the NPT without any solid disarmament proposals from the existing nuclear weapon states. Furthermore, to restrict other states from joining the nuclear club, the 1995 Review Conference urged the state parties to conclude a test ban treaty not later than 1996. For long, in New Delhi's eyes, the NPT symbolized the divide between the nuclear haves and have-nots. An indefinite extension of the NPT legitimized such 'nuclear apartheid' permanently.

This was the context in which Prime Minister Rao ordered the preparation for a nuclear test. This incident, which is now known as India's 'near test', occurred in December 1995 (Ganguly 1999). However, American satellites found the telltale signs of the preparations at the Pokhran test site and diplomatic pressure from Washington followed. The then US ambassador to India, Frank Wisner, urged Rao's principle secretary, A.N. Varma, not to test. Otherwise, India could incur a range of sanctions under the Glenn Amendment. Clinton also called upon Rao to reiterate his ambassador's word, receiving a word from the Indian prime minister that India will not 'act irresponsibly' (Talbott 2004: 37). Under American pressure, Rao demurred. However, by the middle of 1990s, it was clear that the 'relevance of measured ambiguity' in India's nuclear weapons programme and missile capacity 'was coming to an end' (Dixit 1996: 374).

By the time Rao exited the political scene in May 1996, New Delhi openly accepted the imperative to have a nuclear weapons programme for its national security concerns, most evident during its participation in drafting of the CTBT at Geneva. As the draft treaty was being negotiated in the summer of 1996, the nuclear weapon states neither promised any guarantees towards a time-bound disarmament plan nor did they agreed to cap their technological capability in further refining their nuclear arsenals through subcritical testing (MEA 1997: 90–1). India, from the very start of the negotiations, had conditioned its support on these two stipulations. However, for New Delhi, the most troublesome part was the draft treaty's 'entry into force' requirements, which stipulated that 44 specific countries would need to sign and ratify the treaty for its eventual international application. This constituted a clear targeting of India's threshold nuclear status; India, Israel, and Pakistan were the targeted audience. Perceiving its nuclear option being under attack, the Indian envoy to conference on disarmament (CD) on 20 June 1996 declared that 'India would not subscribe to CTBT in its existing form as it was not conceived as a measure towards UND and was not in *India's national security interests*' (emphasis added) (Ghosh 1997: 255). This was the first time in its disarmament diplomacy that India had rejected a treaty solely on the pretext of national security. It also underlined that New Delhi considered maintaining its nuclear option as an imperative. The UNGA voted overwhelmingly in favour of the Australian resolution on CTBT in September 1996. All three conditions put forth by India—linking of the CTBT with the overall goal of nuclear disarmament; its opposition to the 'entry into force' clause of the negotiated treaty; and disallowance of all kinds of nuclear testing, including simulations and zero-yield tests—were sidelined completely (Ghosh 1997). The CTBT discussion also engendered a massive public debate within for the need to go nuclear. Nuclear nationalism was also forcing the hands of Indian decision-makers. The right-wing BJP was the most emphatic supporter of this nuclear nationalism. It had for long supported an open demonstration of India's nuclear capability. Beginning in the early 1990s, BJP had made significant gains in India's political space and in May 1996, it became the single largest party in the Indian Parliament. Under Atal Bihari Vajpayee, for the first time, BJP came to power at the centre.[4]

The question staring at Indian decision-makers was when and how to declare India as a nuclear weapons power? In May 1996, India again verged on a nuclear

[4] For a discussion on BJP's nuclear and foreign policy, see Ogden (2014).

test under the Vajpayee-led BJP government. In fact, as soon as Vajpayee took oath as prime minister, he gave the required authorization to the scientists for conducting a nuclear test. However, in the face of a looming crisis of his government's inability to prove its majority in the Parliament, Vajpayee took a conscious decision to postpone the test, lest the fallout of the events would cripple the next government (Chengappa 2002: 395). In June 1996, H.D. Deve Gowda formed a United Front government, supported by the Indian National Congress from outside. Even when ideologically this particular political dispensation was a left-oriented one, the government continued to support India's nuclear programme by providing much-needed economic assistance.[5] India was not ready to give up its right to nuclear weapons. As the UNGA passed the CTBT in December 1996, the Indian foreign minister stated categorically that 'India will not give up its nuclear weapons option' (FBIS 1996).

The policy enunciated by Rajiv Gandhi was pursued by subsequent political leadership. From Rao till Gujral,[6]

[5] Interview with a senior military officer who was member of the Chiefs of Staff Committee (CoSC), Mumbai, January 2015.

[6] I.K. Gujral was the prime minister of India from April 1997 to March 1998.

all successive prime ministers supported the process of nuclear weaponization and even contemplated nuclear tests. For a wide variety of reasons, however, India maintained its ambivalent nuclear status. This ambivalence was finally broken on May 1998 when India conducted a series of nuclear tests at Pokhran. A nuclear India had finally arrived.

## A Normal Nuclear Power

In March 1998, Vajpayee returned as prime minister to head a BJP-led National Democratic Alliance (NDA) government. Within two months, the prime minister fulfilled BJP's long-held promise of making India a nuclear weapons power when, on 11 May 1998, India conducted three nuclear tests at Pokhran, followed by another couple of tests on 13 May. Twenty-four years after it conducted the PNE in May 1974, India's 'ambivalent' nuclear policy had completed a full circle. If domestic reaction to the tests reflected nationalistic pride, external reactions were of shock and regret. Within a few weeks of India's tests, Pakistan conducted its own nuclear tests. The nuclearization of the subcontinent was now complete. Most of the international community, led by the US, heavily criticized India's decision. Even though India's old strategic partners like France and Russia argued for a

123

cautious approach, the US imposed severe economic sanctions.[7]

By conducting the tests, India had challenged the global nuclear order. In the aftermath, India tried to reconcile its new status with the non-proliferation regime. India based its post-nuclear tests foreign policy on two assumptions. First, the world would understand that nuclear renunciation was not an option for India; India's nuclear weapons were a fait accompli. Second, a nuclear India was too important to be ignored. However, it would first require convincing the global hegemon, the US. Vajpayee, therefore, began a process of genuine reconciliation with the US. Within three months of the test, India and the US agreed to initiate a nuclear dialogue. What ensued was the most comprehensive dialogue process between India and the US in the history of their bilateral relations. The two estranged democracies were now engaged in, to use Jaswant Singh's words, a 'dialogue of the equals'.[8] This process was bolstered immensely by India's

[7] On reactions of major powers to India's nuclear tests in 1998, see Nayyar (2001).

[8] The two interlocutors from the Indian side and the American side were Jaswant Singh and Strobe Talbott. Both Talbott and Singh later delineated upon their experiences in Singh (2006) and Talbott (2004).

responsible nuclear behaviour in the face of Pakistan's blatant provocation during the Kargil War. For the first time after Sino-Soviet clashes in 1969, two nuclear-armed countries were involved in a conventional war. Pakistan was clearly the aggressor; yet, India exercised considerable restraint in not expanding the war beyond the Kargil sector even when it sustained major losses of men and material in the process. India's restraint earned it a recognition of being a responsible nuclear power. It also made the US side fully with India. For the first time since 1971, the US policy in South Asia tilted towards New Delhi. In hindsight, what appeared to be a nuclear flashpoint initially, the Kargil War proved to be a 'paradigm shift' for India–US relations (Mohan 2003: 98).

This responsible nature of India's nuclear behaviour was also displayed in its nuclear doctrine. Four broad principles explained India's nuclear philosophy. First, at the declaratory level, India articulated a vision where nuclear weapons were 'more an instrument of politics rather than a military instrument of war-fighting' (Singh 1998: 11). Nuclear weapons were a political tool geared towards one single objective: to avert the threat of use or actual use of nuclear weapons against it by its adversaries. Second, New Delhi announced that it will follow a policy of 'no first use' (NFU) against nuclear weapon states and of 'non-use' of nuclear weapons against NNWS. Third, India declared its intentions not to enter into an arms race with any of its nuclear adversaries. To this end, it declared a voluntary moratorium on further nuclear testing. Fourth, New Delhi declined to follow the nuclear trajectory of great powers during the Cold War which entailed hair-trigger nuclear alerts and launch-on-warning nuclear posture. Instead, India adopted a purely retaliatory posture which privileged delayed response to a nuclear attack, minimizing the dangers associated with miscalculations and misperceptions. Such nuclear thinking reflected an image of a responsible and restrained nuclear power.

In the aftermath of the tests, India also aligned itself more openly with the non-proliferation regime.[9] For long India had been a 'nuclear outlier'; it now poised itself to support the same non-proliferation regime it had once accused of being discriminatory. India's behaviour was guided by its self-interests of being a nuclear power; or as the government argued in the Parliament, by the 'responsibility and obligation of power' accrued by its nuclear weapons acquisition (Press Information Bureau 1998). New Delhi declared its open 'commitment to non-proliferation'

---

[9] For these changes in Indian policy, see Mohan (2003, 2006).

and took on the responsibility to 'maintain stringent export controls' over its nuclear know-how (Press Information Bureau 1998). It even accepted the logic of the NPT and its foreign minister argued in the Parliament that 'though not a party, India's policies have been consistent with key provision of the NPT that apply to nuclear weapon states' (Press Information Bureau 2000). India also withdrew its reservations to arms control treaties such as the CTBT and the FMCT; being a nuclear power, arms control now became an important policy goal. General and complete disarmament, a long-held Indian position, lost its relative weight in India's nuclear policy. It was easier for India to fight for principles of nuclear disarmament when it was not a nuclear power; harder to live up to them after being one. India had now become a normal nuclear power.

★★★

India's long nuclear journey ended with the tests of 1998. The necessity of having a nuclear deterrent was apparent to Indian decision-makers since the early 1980s. All prime ministers since Indira Gandhi therefore supported the process of nuclearization. Yet, India remained reluctant to openly declare itself as a nuclear weapon state. Only upon a firm rejection of India's appeals for nuclear disarmament in 1988, the then Prime Minister Rajiv Gandhi asked India's nuclear scientists to prepare a deliverable nuclear arsenal. Pakistan's nuclear coercion also made Indian decision-makers focus on the 'software' required for projecting deterrence.[10] Only after the Indo-Pak military crisis of 1990, issues like nuclear command and control took centre stage in Indian nuclear thinking. As the noose of non-proliferation tightened around India with the indefinite extension of the NPT and the CTBT negotiations, Indian policy-makers veered towards nuclear tests. This gradual discourse towards overt nuclearization was given its final shape by Prime Minister Vajpayee in 1998. Twenty-four years

[10] If nuclear weapons and delivery systems are the hardware of nuclear deterrence, institutionalization of nuclear decision-making, development of organizational capabilities, and operational protocols for nuclear use are its software. As Kampani argues, 'Operationalization [of nuclear deterrence] entails the development of soft institutional and organizational routines. It refers to command and control mechanisms, coordination procedures between scientific and military agencies, and training protocols in the military to deploy and explode weapons (stockpile to target sequence). If the weapon systems constitute the hardware, operational routines make up the software that enables use of weapons during war' (Kampani 2014b: 80–1).

after its first nuclear test, India declared itself a nuclear weapon state. Since then, its nuclear trajectory has been extensive; it now stands at the cusp of becoming a major nuclear power. This is elaborated in the next chapter.

# 5

# A Major Nuclear Power

As India crossed the nuclear rubicon, critics prognosticated a grim future for its nuclear forces and its status as a nuclear weapon state. Claims were made that India would remain a 'third tier nuclear state' and 'a low level nuclear power' (Gupta 2001: 1044). It was also forecasted that Pakistan would remain the singular focus of India's nuclear forces and hence, India's nuclear deterrent at best would be confined to the South Asian region. The reasons behind such assumptions were mainly material: shortage of fissile material; technological incapacity to produce reliable delivery systems; ineffective bureaucratic structures (especially of its scientific enclave); and rudimentary command and control systems. Further, the hostile international reaction to India's overt nuclearization threatened to restrict or even punish Indian efforts to further develop its nuclear capabilities. Supporters of India's nuclear tests, on the other hand, made entirely different predictions (Subrahmanyam 1998b). They claimed that since nuclear deterrence required only a few nuclear weapons, India would not enter into an arms race as was the case with great powers during the Cold War. Having managed to enter the 'mainstream paradigm' of nuclear weapon states, India can now 'change the paradigm itself' by forcefully pursuing the quest for nuclear disarmament (Subrahmanyam 1998b: 53). Lastly, mutual nuclear deterrence between Indian and Pakistan would help the two countries achieve peace.

India's nuclear trajectory since Pokhran-II belies all these assumptions. The last two decades have seen India becoming one of the world's major nuclear powers. It has taken rapid strides in technological advancement of its nuclear forces. It has made substantial inroads into the non-proliferation regime and has been accepted as a responsible nuclear power by the international community. It has increasingly behaved like any other nuclear power in history, privileging non-proliferation over nuclear disarmament, so that it retains the exclusivity of power and prestige which accompanies nuclear weapons. Security competition with Pakistan has only intensified and the nuclear equation between the two countries remains highly unstable. Over the years, China has become the real focus of India's nuclear deterrent. India's nuclear trajectory has been remarkably different from what scholars and analysts predicted in 1998. This chapter explains the evolution of India's nuclear policy since 1998, especially the evolving nuclear threat scenario, its emerging nuclear profile, the debate around its nuclear doctrine, its accommodation in the global nuclear order, and finally, its changing attitude on arms control treaties such as the CTBT and the FMCT.

## Pakistan's Nuclear Braggadocio

After the nuclear tests of 1998, dominant political opinion in India claimed that existence of nuclear deterrence would help India–Pakistan attain stability in their bilateral relations. As Prime Minister Vajpayee argued, 'Now both India and Pakistan are in possession of nuclear weapons. There is no alternative but to live in mutual harmony. The nuclear weapon is not an offensive weapon. It is a weapon of self-defense. It is the kind of weapon that helps in preserving the peace' (Karat 1999). Such expectations, however, did not entirely fit Pakistan's grand strategy. In fact, the cover of nuclear weapons provided Pakistan an enabling environment to push further its proxy war in Kashmir and support terrorism in India. Pakistani calculation was simple: the threat of use of nuclear weapons will deter India from launching a conventional war as a strategy to punish Pakistan for its support of terrorism in Kashmir and beyond.

This was first put to test in the summer of 1999, when regular Pakistani troops occupied strategic heights in the Kargil sector of Jammu and Kashmir. Presenting it as a fait accompli, Pakistani army generals believed that India will not opt for a major conventional war as it may escalate to a nuclear one. International pressure to cease hostilities between two nuclear-armed adversaries would further restrict India's choices was another of their assumptions. India, however, responded in force but confined the military action to Kargil, even when voices within the government argued for opening up another front. Pakistan was also quick to issue veiled nuclear threats, creating consternations in the international community which started calling Kashmir as the world's nuclear flashpoint. Eventually India was successful in flushing out the intruders, but this came at a great cost of men and material. If the Indian armed forces made it difficult for Pakistan to sustain its occupation, India's deft diplomacy allowed it to garner support of the international community which labelled Pakistan as the aggressor (Riedel 2009). Restraint shown by New Delhi also earned it the label of a responsible nuclear power. Notwithstanding India's eventual victory, Indian decision-makers learned that Pakistan had become more rather than less risk prone after going nuclear (Government of India 2000: 198). Nuclear weapons provided it a cover to pursue low-level conflict with India.

This strategy was again on display in December 2001 when a group of Pakistan-based terrorists laid siege to the Indian Parliament. The public mood was vociferously in favour of a military response. Under Operation Parakram, Prime Minister Vajpayee did initiate a massive mobilization of Indian armed forces and over half a million troops were deployed over the western border with Pakistan.[1] This coercive military strategy, however, failed because India desisted from initiating a conventional military action against Pakistan. Counter-mobilization by Pakistan, international diplomatic pressure especially from the US, and the threat of nuclear escalation were primarily responsible for India's restraint. The nuclear revolution in the subcontinent had tied Indian strategy in knots.

Even though conventionally stronger, India could not bring to bear its power on Pakistan for the fear of nuclear escalation. India's frustration was rooted in its inability to halt the activities of Pakistan-sponsored

[1] For perceptive discussions of 2001 Parliament attacks and India's military response, see Sood and Shawney (2003).

militant groups in a way that does not threaten major war with potentially nuclear consequences. It was under these circumstances that Indian defence planners started arguing for the possibility of a limited conventional war as the only resort to punish Pakistan for its sub-conventional adventure (Sidhu and Smith 2000). This resulted in a military strategy which some analysts have termed as the 'Cold Start'. First propounded in the Indian Army Doctrine (2004), it involves a rapid cross-border conventional attack to hold limited areas of Pakistani territory for bargaining leverage as a response to Pakistan's continued support to terror.[2] That the limited war doctrine of 'Cold Start' failed to deliver was evident in India's response to the 2008 Mumbai terror attacks. India, once again, desisted from any military action against Pakistan even when sufficient evidence was available of the links between Pakistani intelligence services and the terrorists.

Yet, Pakistan, for its part, has used the pretext of the Cold Start doctrine to expand its nuclear arsenal and to introduce new kinds of nuclear weapons in South Asia.[3] Pakistan now has the world's fastest-

---

[2] A good discussion and analysis of the concept and practice of Cold Start can be found in Ladwig (2008).

[3] A good summary of Pakistani nuclear posture is available in Karl (2015).

expanding nuclear arsenal. It is also further refining and miniaturizing its nuclear warheads and has shifted to plutonium-based nuclear weapons compared to its traditional reliance on uranium. Its missile development programme now boasts a capability of targeting the entire Indian territory, including that of Andaman and Nicobar Islands. To preserve its nuclear forces from an Indian conventional attack or a nuclear strike, Islamabad has now started exploring the possibility of emplacing its nuclear weapons in the high seas. It established a naval strategic force command in 2012 and is believed to have placed nuclear-enabled cruise missiles on its conventional submarines. The most dangerous development, however, has been Pakistan's introduction of tactical nuclear weapons (TNWs) in South Asia. In 2009, Pakistan first tested a short-range nuclear-capable missile called Nasr. With its 60-km range, Nasr is capable of hosting a small-yield TNW. Pakistan has emphatically declared that this weapon system is its response to any conventional thrust by Indian armed forces into its own territory across the international border.

These developments not only suggest an increasing nuclear capability but also reinforce the Pakistani penchant for nuclear risk-taking (Koithara 2003). Its nuclear philosophy conveys the same. Its thresholds for use of nuclear weapons against India are extremely low.

It envisages using nuclear weapons not only in case of a conventional attack but also if New Delhi resorts to economic coercion, such as naval blockade, or foments domestic instability inside Pakistan. To this end, it has now developed a doctrine of full spectrum deterrence: that nuclear weapons will deter all Indian action whether at nuclear, conventional, or sub-conventional level (Tasleem 2016). However, it is the introduction of TNWs in the subcontinent which has been the most destabilizing for nuclear stability on the one hand, and for India's response to continuous provocation by Pakistan on the other. It would require pre-delegation of authority to the military commanders, thereby increasing the dangers associated with misperceptions and miscalculations. There is also a fear that an expanding nuclear arsenal provides more opportunities for nuclear theft by terrorist organizations active within Pakistan, increasing the threat of nuclear terrorism the world over. However, they also pose a peculiar problem for India insofar it creates complexities for India's conventional response to Pakistan's sub-conventional war. As per India's nuclear doctrine, even a low-level nuclear response from Pakistan would invite a massive nuclear strike from India.

In the post-1998 period therefore, Pakistan's nuclear challenge has become only more complex for Indian decision-makers. Its expanding nuclear arsenal, coupled with its risk-prone nuclear strategy, complicates India's deterrence calculus. Pakistan's commitment to proxy wars against India also shows no signs of abating.

## The China Challenge

India is the only nuclear weapon state in the entire world which is surrounded by two hostile nuclear powers. It was the Chinese nuclear test in 1964 that first animated the debate over an Indian nuclear weapons capability. In later decades, China receded into the background of India's concerns and Pakistan became the most immediate motivation for India to go nuclear. However, China did remain a long-term nuclear threat because of the disputed land border, and also because of Beijing's political, economic, and military support to Pakistan (Garver 2001). China's larger strategy in Asia also demanded that India remains embroiled in the South Asian region. A nuclear India was therefore a problem for Beijing.

Notwithstanding the long-term hostility between the two countries, India's nuclear equation with China has remained stable, unlike that of Pakistan. Two reasons were primarily responsible for this muted nuclear competition. First, in the immediate aftermath of the 1998 tests, India's nuclear deterrent was too rudimentary to have been a concern for Chinese decision-makers.

Whereas China was capable of targeting the whole of Indian territory with its medium- and long-range ballistic missiles, India's air-delivered nuclear arsenal faced serious vulnerabilities. Second, Chinese military strategy, unlike that of Pakistan, did not depend upon its nuclear forces. China followed, at least at the level of rhetoric, an 'NFU' policy with regard to the use of nuclear weapons. Being conventionally superior, there was also no need for China to engage in nuclear coercion against India.

However, after the 1998 nuclear tests, India's nuclear trajectory has been determined by a quest to achieve an effective nuclear deterrent against China (Kampani 2013). This drive to attain a stable nuclear balance with China is driven by a number of factors. First, even when China continues to decouple its nuclear forces from its conventional forces, a large gap in nuclear capabilities between India and China may provide the latter with a capability to blackmail India in case of a serious conventional war (Nagal 2015). India–China boundary dispute along the Himalayas continues to dodge any resolution. If anything, China's growing economic and military power in last two decades has made it more and not less adamant of its claims on the Indian territory. Second, after a substantial gap, China has started expanding and modernizing its nuclear forces in response to its growing military competition with the

US.[4] This automatically creates pressure upon India. The most troublesome aspect of Chinese modernization has been its expanding missile capabilities, both conventional and nuclear. Indian military analysts see in this a major missile gap between the two countries as India's missile capabilities have not yet reached that level of sophistication and maturity. This 'strategic imbalance', in the words of a senior Indian military commander, may force India to 'obtain matching or counter capabilities' by boosting the growth of its nuclear deterrent (Nagal 2015: 15–16). Third, the modernization of Chinese nuclear forces also provide her with a counter-force capability: to specifically target India's nuclear forces and infrastructure. These developments in Chinese nuclear infrastructure and thinking engender enormous challenges for India's nuclear deterrent, especially with regard to minimizing the expanding deterrence gap between the two counties.

Sino-Pak nuclear nexus in nuclear proliferation also complicates India's position. China and Pakistan have an extensive record of economic and defence cooperation targeted at complicating India's rise.[5] While

[4] For a perceptive discussion on changing Chinese strategic profile, see Riqiang (2013).

[5] Small (2015) provides a good discussion of Sino-Pak military and strategic alliance.

the two states do not always act in lockstep when it comes to their defence planning and operations vis-à-vis India, they share a multifaceted and strengthening strategic partnership. This has included substantive nuclear proliferation and ambitious infrastructure and defence projects. Pakistan's military and civilian nuclear programmes have long benefited from Chinese technological and economic assistance. Such nuclear cooperation continues to this day. China has also invested in Pakistan's civil nuclear energy programme. Recent developments have included the construction of the Chasma-2 reactor (operational since 2011) and the Chasma-3 and Chasma-4 reactors (under construction since 2011). China is also reportedly helping Pakistan with its nuclear submarine programme. Such extensive nuclear cooperation creates an image that Pakistan and China may at some time coordinate their nuclear strategies to target India. Continued Sino-Pak nuclear nexus may force India to contemplate a two-front nuclear war.

## Expanding Trajectory of India's Nuclear Forces

When India declared itself a nuclear weapon state in 1998, its nuclear capabilities were rudimentary.[6] The only

[6] For an extensive discussion on India's changing nuclear profile in the last two decades, see Joshi et al. (2016).

operational part of its nuclear forces was the air vector based upon a single platform, the Mirage fighter aircraft. Its missile capabilities were growing with three different missiles systems, namely, Prithvi, Agni-1, and Agni-2, that had been at least test-fired but were far from ready to carry India's nuclear arsenal. A nuclear submarine programme was initiated in the late 1970s but the progress had been haltingly slow. Over the last two decades, however, India's nuclear forces have expanded dramatically.

Today, India has an extensive fleet of fighter aircraft which can deliver a nuclear payload. Its missile programme has grown substantially with the Agni series of missiles becoming the primary delivery vehicle. With the coming of Agni-5, India now has an intercontinental ballistic missile capability. These missile platforms have been silo-hardened and made road and rail mobile. India is also working on developing the multiple independently targetable re-entry vehicle (MIRV) technology which will allow its missiles to carry multiple nuclear warheads, increasing the bang for the buck of its nuclear arsenal. The nuclear submarine programme has achieved fruition with the commissioning of the first SSBN,[7] *INS Arihant*. A series of nuclear submarines are now under production.

[7] An SSBN is a nuclear-powered submarine capable of launching ballistic missiles armed with nuclear weapons. An

The trajectory of India's nuclear forces in the last two decades has been mainly determined by two factors: credibility of its second-strike capability and the evolving nuclear situation in its neighbourhood. Deterring its adversaries from the use or the threat of use of nuclear weapons requires a credible secure second-strike capability. An effective second-strike capability is only ensured when India's nuclear forces can withstand an initial onslaught of nuclear attack from the enemy and still a formidable force survives for nuclear retaliation.

This singular requirement has been the most important driver of India's nuclear force developments. The air vector was the first leg of India's nuclear delivery. India's fleet of fighter aircraft which can deliver a nuclear payload has expanded from Mirage fighter aircraft in 1998 to Sukhoi 30-MKIs and Jaguars IS/IB fighter aircraft. The acquisition of Rafael multi-role combat aircraft may also be added to the mix. Yet, this is also the most vulnerable vector of nuclear delivery to pre-emptive attacks by the adversary. It is also highly vulnerable to massive attrition by sophisticated air defence systems while performing a nuclear mission.

---

SSN, on the other hand, is propelled by nuclear power but is armed with conventional weapons and is generally known as an attack nuclear submarine.

Ballistic missiles, therefore, have always remained the most preferred form of nuclear delivery. For one, their effective range can far outperform the capability of fighter aircraft. They can carry bigger payloads too. Most importantly, once silo-hardened and made rail or road mobile, they are very difficult to track and target. India's missile programme, therefore, has been the most important venue of its expanding nuclear delivery capabilities. From hosting just three platforms in 1998—Prithvi, Agni-1, and Agni-2—limited in range to targets in Pakistan, today New Delhi is building Agni-5, able to reach all targets in China, and working on Agni-6, intended to extend even further. To further pose a sign of robust intent against potential Chinese aggression, Agni-5 and Agni-6 are also being designed to host MIRV warheads, increasing their destructive capacity. Indian missile developments are thus reaching new heights of technical maturity. However, Prithvi, Agni-1, and Agni-2 are the only missiles that have actually been inducted into India's nuclear forces; rest are still in the R&D stage. This illustrates that full integration of the later Agni platforms into India's nuclear force is still a future aspiration than a technical reality.

The most ambitious and path-breaking addition to India's nuclear arsenal, however, has been the commissioning of India's first SSBN, *INS Arihant.*[8] India's Draft Nuclear Doctrine, released in 1999 by the National Security Advisory Board (NSAB), had stated the requirement of a nuclear triad for effective projection of India's nuclear deterrent. If land-based missiles and aircraft delivery systems constitute the first two legs of the nuclear triad, an underwater nuclear launch capability mounted on nuclear submarines forms the third. The sea-based nuclear launch capability is important as it can provide for a 'post-surprise attack-survivable force' and hence, would have a 'deep stabilising effect' on nuclear deterrence in the region (Menon 2000: 225–6). In the Indian case, there are two main reasons behind this thrust upon developing a force of nuclear-armed submarines or SSBNs, with the concomitant armament of long-range ballistic missiles tipped with nuclear warheads (Tellis 2001). First, the removal of India's nuclear forces from the mainland would effectively neutralize the enemy's nuclear targeting. More nuclear weapons at sea would automatically mean less vulnerability. Second, a naval nuclear force based on nuclear submarines is very

[8] For a discussion on the impact of an SSBN on India's nuclear capability and its strategic environment, see Joshi and O'Donnell (2014).

difficult to detect and destroy, which provides for unhurried retribution after careful evaluation though ensuring that the effect on the enemy would be catastrophic. These characteristics of a sea-based nuclear force complement India's nuclear doctrine, which calls for NFU and massive retaliation. It is therefore that SSBNs in the future may become the primary vector of India's nuclear force projection. Yet, this leg of India's nuclear deterrent will take some time before becoming fully operational. The *Arihant* was launched in 2009 and commissioned in 2016, but it must be seen largely as a technology demonstrator. A series of nuclear submarines with major R&D modifications are now being constructed. These SSBNs will not only be more powerful but would carry longer-range missiles, providing it an ultimate nuclear *force de frappe*.[9]

If acquiring an invulnerable second strike is the primary driver of India's nuclear force development, the evolving nuclear situation in its neighbourhood, as has been discussed in the previous sections on Pakistan and China, also demands constant reappraisal of India's capabilities and future requirements so that its nuclear deterrent remains potent. This requirement

[9] Interview with a former director of the Advanced Technology Vessel (ATV) project in New Delhi, 25 March 2015.

explains many of India's R&D programmes, which may not strictly fit with the traditional requirements of a purely retaliatory deterrence force. For example, since 2003, India has invested heavily in developing a ballistic missile defence system. A number of factors in recent times have made India interested in these systems: China and Pakistan's growing nuclear arsenals; chances of accidental launch from Pakistan; fears of a bolt-from-the-blue strike on India; and Pakistan's unstable political situation and the growing influence of non-state actors, especially in Pakistani body politic (Nagal 2016).

This expanding nuclear force has led Indian decision-makers to critically think about the command and control of India's nuclear arsenal. Following the 1998 nuclear tests, the Indian government established new institutions to manage nuclear policy-making. This replaced a previous informal command chain concentrating authority in the PMO. In 2003, the Nuclear Command Authority (NCA) and Strategic Forces Command (SFC) were established to manage the command and control of nuclear forces.[10] The NCA exists for civilian officials to consider and issue nuclear deployment orders. The membership of the

[10] For a discussion of the evolving command and control structures, see Koithara (2012).

NCA consists of two tiers: a political council chaired by the prime minister and an executive council chaired by the national security advisor. The second institution is the SFC. This is headed by a rotating military chair and the three service chiefs, and exists to organize the military execution of orders emanating from the NCA. In the event of a decision to use nuclear weapons, the command structure is so organized that the decision is made solely by the NCA political council; its execution organized by the NCA executive council; and ultimate execution conducted by the SFC (Pant 2007).

The existence of civil–military coordination mechanisms within the NCA structure, as well as the two dedicated planning groups detailed earlier, are evidence that the Indian government has achieved more effective political and technical oversight over nuclear force development as compared to previous eras. The structural participation of the military in these multiple elements of the nuclear command chain, in this reading, suggests its greater involvement in nuclear force development (Kampani 2014a).

## The Problems of Signalling Deterrence

Following its nuclear tests of May 1998, India decided upon a doctrinal formulation of 'credible minimum deterrence' (CMD). In August 1999, the NSAB issued

India's Draft Nuclear Doctrine. India's doctrine consists of three broad principles. First, at the declaratory level, India has articulated a vision where nuclear weapons are political instruments rather than effective weapons of war (Singh 1998: 11). Second, India adheres to a policy of NFU of nuclear weapons. For Indian decision-makers, for both political and military reasons, NFU has appeared to be a risk worth taking. The NFU comported well with India's overall nuclear philosophy that nuclear weapons should never be used in the battlefield. Since nuclear weapons were only for deterrence, any first use of nuclear weapons was out of question. The third important aspect of the ideological component of New Delhi's nuclear deterrence is, therefore, centred on India's responses to the threat of use of nuclear weapons by its adversaries or actual use of nuclear weapons in case deterrence breaks down. India's nuclear doctrine has maintained an 'assured retaliation' posture. The posture of assured retaliation is based on the premise that deterrence works on the logic of punishment: the threat of retaliation maintains deterrence.

In the period between the 1998 tests and the declaration of India's official nuclear doctrine on 4 January 2003, the definitions of NFU and the posture of 'assured retaliation' underwent some shifts. From a strict NFU policy in 1999, India had, by 2003, conditioned its NFU pledge by declaring that it may retain the right to respond with nuclear weapons in case its territory or its armed forces anywhere in the world were attacked by chemical or biological weapons. On the issue of quantum of punishment, India's retaliatory strategy moved towards a more muscular approach. The volume of retaliation took an ascendant trajectory: from 'punitive retaliation' in August 1999 to 'massive retaliation' in January 2003. However, there have been no further official revisions since then, and Indian analysts argue that both NFU and 'assured retaliation' continue to define India's nuclear doctrine.

In the recent past, four developments in India's external and internal environment have led to a renewed debate around India's nuclear doctrine. First, Indian nuclear and conventional strategies have not been able to adequately answer the challenge of Pakistan-sponsored sub-conventional warfare in the subcontinent. Second, the increasing volume and sophistication of Pakistan's nuclear arsenal, and especially its development of TNWs, has created doubts in Indian strategic circles regarding the credibility of New Delhi's nuclear deterrent. Third, Chinese nuclear force modernization has also generated additional pressure on India's nuclear forces as seen within India, increasing the perceived deterrence gap between the two. Finally, India's own growing strategic capabilities are challenging the doctrine (Narang 2013).

All these factors—the increasing lethality and range of Pakistan's arsenal; India's inability to resolve the Pakistan conventional–nuclear dilemma; China's nuclear modernization; and the growing sophistication of India's nuclear capabilities—have ignited a domestic debate in India over the need to revise the doctrine. The two most important points of current discussion around the nuclear doctrine are India's NFU pledge and its policy of massive retaliation. These precepts were fundamental to India's nuclear thinking when the doctrinal plans were first conceived in the post-1998 period. In the current strategic churning over nuclear doctrine, they are also the most debated.

That NFU has assisted India in projecting itself as a responsible and restrained nuclear power is accepted by most Indian analysts. Being essentially a defensive policy, NFU has helped in 'reassuring globally that India is not an aggressive power' (Nagal 2014: 13). However, as critics now argue, the challenges posed by the evolving strategic situation far outweigh the soft power benefits accrued by the 'passivity' of the NFU pledge. First, an NFU pledge allows the adversary to carry out 'large scale destruction' even before a massive retaliation can be launched against it. Second, an NFU policy also restricts India's military options; it cannot attrite the enemy's strategic assets through selective counter-strike targeting of its nuclear forces.

If NFU is being questioned, so is India's strategy of retaliating massively to any form of nuclear attack. In the official pronouncement of the nuclear doctrine in 2003, India postulated that its response to any kind of nuclear attack would be 'massive'. Critics now argue that this certitude of massive response suffers from huge credibility problems. The most likely use of nuclear weapons in South Asia pertains to the scenario of Pakistan availing its TNWs against Indian armed forces. Such low-level nuclear use, even when deemed as first use of nuclear weapons, cannot believably invite massive retaliation from India. An additional but related issue with the doctrine of massive retaliation is the issue of proportionality of use of force. To threaten extinction of the enemy, which is inherent in the policy of massive retaliation, against low-yield, local use of battlefield nuclear weapons goes against the logic of proportionality of response. However, the issue of political will is also a key concern: would Indian decision-makers be ready to walk the talk in case the adversary resorts to nuclear first use? The issue of political resolve is particularly problematic for a policy of 'massive retaliation' because most critics believe that Indian political class is highly risk-averse. This averseness to political risk was manifest in India's response to crisis situations in the past, whether it was the Kargil War, the Parliament attack and military mobilization crisis of 2001–2, or the more recent Mumbai attacks.

There has been pressure building upon Indian decision-makers to review India's nuclear doctrine. For one, there is some argument towards diluting the NFU. Pre-emption of Pakistan's use of TNWs is gaining ground among those who advocate a first-use policy. Former Indian National Security Advisor Shivshankar Menon has argued that one 'potential grey area' where India could resort to pre-emption is where New Delhi is 'certain that adversary's launch (of nuclear weapons) was imminent' (Menon 2016: 164). Similarly, on massive retaliation, sceptics have argued for a number of other options. The common thinking behind these options is to settle for less than punishing Pakistan 'massively' for its battlefield use of nuclear weapons. The need, therefore, is to dilute the quantum of punishment in the doctrine to the earlier posture of 'punitive retaliation'. This may provide the Indian decision-makers much-needed flexibility to deal with Pakistan's low-level use of battlefield nuclear weapons. However, even critics of massive retaliation admit that adopting a more flexible retaliatory doctrine at this stage may send 'wrong signals' to Pakistan and other adversaries. However, most importantly, a flexible response posture also does not guarantee India its most important objective: to deter Pakistan's use of nuclear weapons in the first place. India is, therefore, caught between a rock and a hard place when it comes to its retaliatory posture.

India's nuclear doctrine, as propounded in 1999 and revised in 2003, signalled India's commitment to being a responsible nuclear power. Its status quo bias where nuclear weapons were considered as only a deterrent against threat or use of nuclear weapons, however, has appeared to be ineffective against risk-prone states like Pakistan. Both of its fundamental premises of the NFU and 'retaliating massively' to a nuclear strike are now being questioned. However, successive Indian leadership has affirmed its faith in the original precepts of its nuclear philosophy. In fact, three different prime ministers, including the government of Narendra Modi, have affirmed the same doctrinal principles over a period of 20 years. The doctrine has tolerated the pressure of the changing security environment, organizational impulses within the military, technological sophistication of its arsenal, growing discontent among the strategic elites on the existing doctrine, and also the change of government at the centre.

However, the fundamental problem for Indian decision-makers and strategists is whether a change in doctrine would really achieve India's main objective. one of deterring nuclear use by the adversary. Indian policy-makers remain yet to be convinced that a dilution of the NFU and a flexible retaliatory strategy would help achieve that end. This tension will, therefore, continue to bedevil India's nuclear policy. This makes it imperative for the government to officially conduct periodic reviews of the nuclear doctrine, a practice which must be institutionalized as is the case with other nuclear weapon states.

## Global Nuclear Order

When in 1998 India conducted a series of nuclear tests, the global opinion was stacked against New Delhi for having posed one of the biggest challenges to the non-proliferation regime. However, within a decade of the nuclear tests, India obtained de facto recognition as a nuclear weapon state with the successful conclusion of India–US civilian nuclear agreement in 2008. It was recognized as a 'state with advanced nuclear technology' and the technological barriers emanating out of India's difficult relationship with the non-proliferation regime were also removed (MEA 2005). The NSG, originally conceived to target India's nuclear programme, unanimously granted an exemption to New Delhi to trade in sensitive nuclear technologies and materials.

This turnaround in India's fortunes was a result of two important strategic developments after 1998 (Mohan 2006). First, India's economic rise with high growth rates provided a geo-economic heft to Indian diplomacy. Accommodation of this rising power therefore fitted well with historical trends in international politics. Second, for the US, despite its position as the chief architect and enforcer of the non-proliferation regime, India was a natural partner in checking the strategic rise of China as Washington's most probable future challenger. These strategic reasons for Indian accommodation were complemented by India's non-proliferation policy approaches. India changed its historic resistance to the NPT and came out in full support of the principle of non-proliferation. Second, India agreed to adhere to a strict export control policy. In June 2005, India enacted the Weapons of Mass Destruction and their Delivery Systems (Prevention of Unlawful Activities) Act, also known as the WMD Act. This legislation was part of India's commitment to establish domestic controls to prevent the proliferation of nuclear, chemical, or biological weapons and their means of delivery. Moreover, India has unequivocally sided with the global community's efforts to restrict further proliferation, whether it was the case of Iran or Syria.

These shifts in India's non-proliferation policy did indeed help its quest for accommodation in the international nuclear order. When the US president declared in June 2005 that 'as a responsible state with advanced nuclear technology, India should acquire the same benefits and advantages as other states', India's track record on nuclear non-proliferation was held to be one of the benchmarks of its responsible nature (MEA 2005). This also started the process of India–US civilian nuclear agreement. For three long years beginning in 2005, India and the US negotiated India's accommodation in the nuclear non-proliferation regime (Pant 2011). India agreed to structurally separate its nuclear weapons programme from its civilian nuclear energy programme, sign an IAEA Additional Protocol, give guarantees to continue its nuclear test moratorium, and negotiate an FMCT. The US, for its part, changed its domestic nuclear laws to restart nuclear technology trade with India, and also promised to secure an Indian exemption from NSG nuclear technology export restrictions. In September 2008, the NSG unanimously agreed to permit India to engage in nuclear trade, leading to the civilian nuclear deal getting final approval from the US President George W. Bush.

The NSG waiver was a historic landmark in India's relations with the non-proliferation regime. However, the civilian nuclear agreement with the US constituted only a partial accommodation. Post-2008, the debate within India veered towards gaining full accommodation in the international nuclear order as India was not a member of any of the major technology denial regimes, such as the NSG, MTCR, or the Australia Group. India is still seeking admission to the NSG as a full member, but has obtained membership of the MTCR, Australia Group, and Wassenaar Arrangement groupings regulating export of missile, chemical/biological, and sensitive conventional technologies, respectively. India's candidacy for the NSG has received support from all major powers, including the US, Australia, France, Germany, Russia, South Korea, and the UK. However, India's NSG challenges remain substantial. First, NSG membership is agreed upon by group consensus, and Indian diplomacy has not yet been able to overcome the continuing opposition by several members, including Ireland, the Netherlands, and Switzerland. India is unwilling to yield to demands for stronger non-proliferation commitments, which could include adopting a permanent test ban or ending fissile material production. The impasse on non-proliferation commitments notwithstanding, the biggest obstacle to India's NSG membership is China. Unlike other major powers, China has not been particularly enthusiastic about accepting India as a nuclear weapon state. China viewed the Indo-US nuclear deal with hostility: an attempt by the US to prop up India as a challenger to

China's hegemony in Asia. China's strategy to effectively sabotage India's NSG membership is by advocating a quid pro quo for Pakistan. Given Islamabad's past problems with proliferation, that is, the A.Q. Khan network, such a proposal has hardly any takers in the NSG. Its strategic merit, however, lies somewhere else: by linking India's membership with Pakistan, it not only invokes the fear of a crumbling nuclear regime under the weight of the exceptions being granted to India, as many committed non-proliferationists argue, but it also takes care of China's all-weather friendship with Pakistan.

## Disarmament and Arms Control

The major issue in arms control which has had huge consequences on India's nuclear policy is its attitude to the CTBT and the FMCT. As we have seen in the previous chapters, just like the non-proliferation regime, India's relationship with these arms control measures has a highly chequered history. However, the nuclear weapons tests of 1998 paved the way for a qualitative shift in India's posture on both of these arms control measures. Soon after, India declared a unilateral moratorium on further nuclear testing. India also showed certain accommodation towards the CTBT. A number of reasons explained this shift. First, India had achieved its primary security interests by exploding a series of nuclear devices. Second, a positive approach towards CTBT could also help New Delhi end its diplomatic isolation in international community. Third, there was also enormous pressure from the US for India to sign the treaty. All these factors contributed to the change in policy insofar that India now intends to wait and watch the global developments on arms control rather than actively scuttle them. Yet, this wait and watch policy is also informed by some debates within over the merits and demerits of agreeing to treaties like the CTBT and FMCT. The most significant domestic opposition to the CTBT emanates out of the controversy regarding the yield of the 1998 test. In recent years, doubts around the 1998 nuclear tests have been rekindled by testimonies of some senior nuclear scientists. In August 2009, K. Santhanam, project director of the 1998 tests, publicly declared that the thermonuclear device was a 'fizzle', or that it had unexpectedly underperformed compared to its planned yield. If in future India decides to set limits to its nuclear programme, warhead yields would be the first major determinant. Second, new delivery systems such as the sea-launched ballistic missiles (SLBMs) would require further miniaturization of nuclear warheads. Lastly, the confidence of the military in India's warhead capabilities would be equally consequential, which some have argued has been rather low. All these reasons have resulted in certain internal opposition to the CTBT.

As has been the case with the CTBT, India's position on the FMCT, in the post-1998 period, has seen a positive change. After the 1998 nuclear tests, the Vajpayee government indicated its willingness to negotiate an early conclusion of the FMCT (Talbott 2004). In 1999, India also dropped its reservation on effective, multilateral, and intrusive verification mechanism under the FMCT, which it had earlier insisted upon. Therefore, by the end of 2000, India's position on the FMCT was aligned with that of the US and other nuclear weapon states. These changes were also evident during the negotiations on Indo-US civilian nuclear agreement. As a quid pro quo to the easing of US restrictions on nuclear trade, India committed itself on 'working with the United States for the conclusion of a multilateral Fissile Material Cut-off Treaty' (MEA 2005). However, unlike other P-5 countries[11] with the exception of China, India had declined to observe a unilateral moratorium on further production of fissile material. Such changes in

[11] P-5 refers to the UN Security Council's five permanent members, namely, China, France, Russia, the UK, and the US.

attitude notwithstanding, the critical factor in India's future policy on the FMCT would be defined by its deterrence requirements: how much fissile material would be needed to assure a credible deterrent posture? While some minimalists in India's strategic community have argued for a total of around 100 nuclear warheads as providing India with sufficient capability, some maximalists have argued for a force strength of over 400 weapons.

Therefore, India's national security interests will dictate its eventual decision on arms control issues such as the CTBT and the FMCT. These trends will continue until there is a larger movement towards nuclear disarmament globally. Under the current geopolitical scenario, where major nuclear powers such as the US, Russia, and China are laying more emphasis on their nuclear arsenals, any movement towards disarmament is hard to imagine. Lack of any major disarmament initiative on India's part also signifies that its position on disarmament and arms control has come much closer to those of other nuclear weapon states.

★★★

India's nuclear policy has been driven by its quest for both security and status. The fear of China, and more so Pakistan, pushed India to cross the nuclear rubicon.

Its interest in emerging as a major global power has always been a factor in its nuclear diplomacy: it first championed the cause of nuclear disarmament and having gone nuclear, today professes allegiance to the same global nuclear order it once called described as akin to 'nuclear apartheid'. Consideration of prestige, on the other hand, did not allow India to accept the status of an NNWS and to submit to pressures and sanctions imposed by world's nuclear elites. Today, it aspires for complete accommodation in the global nuclear order. Security and status will continue to guide India's nuclear policy in future.

Pakistan and China will remain the two most important challenges to India's nuclear policy, especially as they represent two distinct kinds of nuclear threats. With China, India does not face a risk-prone nuclear power, but the gap in capabilities must be reduced. For this, India would need to fully operationalize its intercontinental ballistic missiles and the SSBNs. Deterrence parity with China will help achieve stable deterrence. However, India would also need to beef up its conventional military power. Huge asymmetries in conventional forces would increase the pressure upon New Delhi to resort to nuclear first use. If India wants to maintain the present distinction between its conventional and nuclear strategy, this is a must. With Pakistan, India possesses an adequate nuclear deterrence capability. However, the problem exists in deterring Pakistan from its sub-conventional pinpricking. The Indian decision-makers and strategists have struggled to find an answer to this problem. How India responds to this unique nuclear challenge is only a matter of conjunction but it will continue to be a key factor in India's strategic thinking.

India's interests will also make it behave more and more as a 'normal' nuclear power. The revisionist agenda of its nuclear policies during the Cold War reflected its self-interests. Its post-1998 nuclear agenda reflects its comfort with the existing status quo. Non-proliferation has become a major policy objective and India has been underscoring its responsible credentials in curbing proliferation. Having gatecrashed into the club of nuclear elites, it now wants to maintain the exclusivity of the privilege and power associated with nuclear weapons. The agenda of disarmament has receded to the background, even though India may periodically trumpet its inclination for complete elimination of nuclear weapons. Its national security interests have overpowered its past moralistic approach to the problems accompanying nuclear weapons.

Finally, nuclear issue has been a major factor in India's quest for a great power status from the very beginning. Nehru not only saw in the atom a symbol of modernity and technological progress but also understood the power which nuclear weapons bestowed upon their possessors. As nuclear weapons became central to great power identity during the Cold War, India refused to let go its strategic autonomy to possess nuclear weapons and to be counted as a great power. It is not a mere coincidence that India's rise in the global order in the last two decades coincided with its rise as a nuclear weapons power. In the atomic age, nuclear weapons have been the prime symbol of status and India's ambitions on the global stage require no less. This also means that India will not adopt unilateral measures to cut down on its nuclear profile until and unless there is larger trend towards delegitimization of nuclear weapons worldwide. They will continue to remain central to its identity as a major power and to its quest for emerging as a great power.

# Bibliography

Abraham, I. 1998. *The Making of Indian Atomic Bomb: Science, Secrecy and the Postcolonial State*. London: Zed Books.

Andersen, R. 1975. 'Building Scientific Institutions in India: Saha and Bhabha', Occasional Paper Series No. 11, Centre for Developing Area Studies, McGill University, Montreal.

———. 2010. *Nucleus and the Nation: Scientists, International Networks and Power in India*. Chicago: University of Chicago Press.

Babu, B.R. 1968. 'Nuclear Proliferation and Stability in Asia', *Economic and Political Weekly*, 3(36): 1365–8.

Bader, W. 1968. *The United States and the Spread of Nuclear Weapons*. New York: Pegasus.

Basrur, R. 2006. *Minimum Deterrence and India's National Security*. Stanford, CA: Stanford University Press.

Bhabha, H.J. 1956. 'General Plan for Atomic Energy Development in India', in Department of Atomic Energy (ed.), *Proceedings of the Conference on the Development of Atomic Energy for Peaceful Purposes in India*, New Delhi, 26–27 November, 1954, pp. 7–20. Bombay: Government of India.

Bhatia, S. 1979. *India's Nuclear Bomb*. New Delhi: Vikas Publishing House.

Bhutto, Z.A. 1979. *'If I am Assassinated…'*. New Delhi:Vikas Publishing House.

Brecher, M. 1966. *Succession in India: A Study in Decision-Making*. London: Oxford University Press.

———. 1968. *India and World Politics: Krishna Menon's View of the World*. London: Oxford University Press.

Buchan, A. 1965. 'The Security of India', *The World Today*, 21(5): 210–17.

Central Intelligence Agency (CIA). 1981. 'India's Reactions to Nuclear Developments in Pakistan', *CIA FOIA Online Reading Room*, SNIE 32/38-I. Available at https://www.cia.gov/library/readingroom/docs/DOC_0005403744.pdf, accessed on 10 January 2018.

Chakma, B. 2004. *Strategic Dynamics and Nuclear Weapons Proliferation in South Asia: A Historical Analysis*. Brussels: Peter Lang.

Chari, P.R. 1978. 'An Indian Reaction to U.S. Nonproliferation Policy', *International Security*, 3(2): 57–61.

Charpie, R. 1955. 'The Geneva Conference', *Scientific American*, 193(4): 27–33.

Chengappa, R. 2002. *Weapons of Peace:The Secret Story of India's Quest to be a Nuclear Power*. New Delhi: HarperCollins.

Coffey, J. 1971. 'Nuclear Guarantees and Nonproliferation', *International Organization*, 25(4): 836–44.

Department of Atomic Energy (DAE). 1970. *Nuclear Weapons: A Compilation Prepared by the Department of Atomic Energy*. Bombay: Government of India.

Deshmukh, C. 2003. *Homi Jehangir Bhabha*. New Delhi: National Book Trust.

Divine, R.A. 1981. *Blowing on the Wind: The Nuclear Test Ban Debate, 1954–1960*. Oxford: Oxford University Press.

Dixit, J.N. 1995. *Anatomy of a Flawed Inheritance: Indo-Pak Relations 1970–1994*. New Delhi: Konark.

———. 1996. *My South Block Years: Memoirs of a Foreign Secretary*. New Delhi: UBS Publishers.

———. 2002. *An Afghan Diary: A New Beginning*. New Delhi: Konark.

Doctor, A.H. 1971. 'India's Nuclear Policy', *The Indian Journal of Political Science*, 32(3): 349–56.

Edwardes, M. 1965. 'Illusion and Reality in India's Foreign Policy', *International Affairs*, 41(1): 48–58.

Erdman, H.L. 1967. *The Swatantra Party and Indian Conservatism*. Cambridge: Cambridge University Press.

Foreign Information Broadcast Service (South Asia). 1996. 'India: Gujral says that Indians Cannot give up Nuclear Option', FBIS-TAC-97-002, 2 December.

Gandhi, I. 1972. 'India and the World', *Foreign Affairs*, 51(1): 65–77.

Gandhi, R. 1988. 'A World Free of Nuclear Weapons: Speech at the United Nations General Assembly, 9 June 1988', in *India and Disarmament: An Anthology*, pp. 280–94. New Delhi: Ministry of External Affairs, Government of India.

Ganguly, S. 1999. 'India's Pathway to Pokhran II: The Prospects and Sources of India's Nuclear Weapons Test', *International Security*, 23(4): 148–77.

Garver, J.W. 2001. *Protracted Contest: Sino-Indian Rivalry in the Twentieth Century*. Seattle and London: University of Washington Press.

Gavin, F. 2012. *Nuclear Statecraft: History and Strategy in America's Atomic Age*. Ithaca: Cornell University Press.

Ghosh, A. 1997. 'Negotiating the CTBT: India's Security Concerns and Nuclear Disarmament', *Journal of International Affairs*, 51(1): 239–61.

Goldschmidt, B. 1982. *The Atomic Complex: A Worldwide Political history of Nuclear Energy*. Illinois: American Nuclear Society.

Goldsmith, B. 1986. 'A Forerunner of the NPT? The Soviet Proposals of 1947', *IAEA Bulletin*, 28(1): 58–64.

Gordon, S. 1994. 'Capping South Asia's Nuclear Weapons Programme: A Window of Opportunity', *Asian Survey*, 34(7): 662–73.

Government of India. 2000. *From Surprise to Victory: Kargil Review Committee Report*. New Delhi: Sage.

Greenberg, D.S. 1962. 'Nuclear Energy for India: US Positions on Safeguards Raises Concern of IAEA', *Science*, 138(3546): 1245–6.

Gupta, A. 2001. 'India's Third-tier Nuclear Dilemma: N Plus 20?', *Asian Survey*, 41(6): 1044–63.

Gupta, B.S. 1978. 'Dilemma with Anguish: India, Morarji and the Bomb', T.T. Poulose (ed.), *Perspectives of India's Nuclear Policy*, pp. 224–39. New Delhi: Young Asia Publications.

———. 1983. *Nuclear Weapons? Policy Options for India*. New Delhi: Sage.

Halperin, M.H. 1965. 'Chinese Nuclear Strategy: The Early Post-Detonation Period', *Asian Survey*, 5(6): 271–9.

Halsted, T. 1974. 'The Spread of Nuclear Weapons: Is the Dam about to Burst?', *Arms Control Today*, 4(11): 1–4.

Haskins, C.P. 1946. 'Atomic Power and American Foreign Policy', *Foreign Affairs*, 24(4): 591–609.

Hecht, G. 1998. *The Radiance of France: Nuclear Power and National Identity after World War II*. Cambridge: MIT Press.

Holloway, D. 1994. *Stalin and the Bomb: The Soviet Union and Atomic Energy, 1939–1956*. Yale: Yale University Press.

Indian Army. 2004. *Indian Army Doctrine*. Shimla: Indian Army Training Command Headquarters, October. Available at https://www.files.ethz.ch/isn/157030/India%202004.pdf, accessed on 10 April 2018.

Jayagopal, J. 1978. 'India's Nuclear Policy and Pakistan's Nuclear Responses', in T.T. Poulose (ed.), *Perspectives of India's Nuclear Policy*, pp. 189–204. New Delhi: Young Asia Publications.

Jayaprakash, N.D. 2000. 'Nuclear Disarmament and India', *Economic and Political Weekly*, 35(7): 525–33.

Joshi, Y. 2015. 'The Imagined Arsenal: India's Nuclear Decision-making, 1973–76', Working Paper No. 6,

Nuclear Proliferation International History Project, Woodrow Wilson Center.

―――. 2017a. 'Debating the Nuclear Legacy of India and One of Its Great Cold War Strategist', *War on the Rocks*. Available at https://warontherocks.com/2017/03/debating-the-nuclear-legacy-of-india-and-one-of-its-great-cold-war-strategists/, accessed on 10 January 2018.

―――. 2017b. 'Waiting for the Bomb: PN Haksar and India's Nuclear Policy in 1960s', Working Paper No. 10, Nuclear Proliferation International History Project, Woodrow Wilson Center.

―――. 2018. 'Between Principles and Pragmatism: India and the Nuclear Non-proliferation Regime in the Post-PNE Era, 1974–1980', *International History Review*, 2 January (Electronic Version), https://www.tandfonline.com/eprint/eJm682iAkfiCPTBGP3qS/full.

Joshi, Y. and F. O'Donnell. 2014. 'Lost at Sea: The *Arihant* in India's Quest for a Grand Strategy', *Comparative Strategy*, 33(5): 466–81.

Joshi, Y., F. O'Donnell, and H.V. Pant. 2016. *India's Evolving Nuclear Force and Implications for US Strategy in the Asia-Pacific*. Carlisle, Pennsylvania: Strategic Studies Institute, US Army War College.

Kampani, G. 2013. 'India: The Challenges of Nuclear Operationalization and Strategic Stability', in A. Tellis, A. Denmark, and T. Tanner (eds), *Strategic Asia 2013–14: Asia in the Second Nuclear Age*, pp. 99–128. Seattle: National Bureau of Asian Research.

————. 2014a. 'Is the Indian Nuclear Tiger Changing its Stripes?', *The Nonproliferation Review*, 21(3–4): 386–7.

————. 2014b. 'New Delhi's Long Nuclear Journey: How Secrecy and Institutional Roadblocks delayed India's Weaponization', *International Security*, 38(4): 71–114.

Kapur, A. 1978. 'The Canada–India Nuclear Negotiations: Some Hypotheses and Lessons', *The World Today*, 34(8): 311–20.

————. 1979. *International Nuclear Proliferation: Multilateral Diplomacy and Regional Aspects*. London: Praeger.

Karat, P. 1999. 'Kargil and Beyond', *Frontline*, 16(14). Available at http://www.frontline.in/static/html/fl1614/16140150.htm, accessed on 10 January 2018.

Karl, D. 2015. 'Pakistan's Evolving Nuclear Posture: Implications for Deterrence Stability', *The Nonproliferation Review*, 21(3–4): 317–36.

Karnad, B. 2005. *Nuclear Weapons and Indian Security: The Realist Foundations of Strategy*. New Delhi: Macmillan India.

Kasturi, B. 1995. *Intelligence Services: Analysis, Organisation and Function*. New Delhi: Lancer.

Keeley, J.F. 1980. 'Canadian Nuclear Export Policy and the Problems of Proliferation', *Canadian Public Policy*, 6(4): 614–62.

Kennedy, A. 2011. 'India's Nuclear Odyssey: Implicit Umbrellas, Diplomatic Disappointments and the Bomb', *International Security*, 36(2): 120–152.

Khan, F.H. 2014. *Eating Grass: The Making of the Pakistani Bomb*. New Delhi: Cambridge University Press.

Kishore, M.A. 1969. *Jana Sangh and India's Foreign Policy*. New Delhi: Associated Publishing House.

Koithara, V. 2003. 'Coercion and Risk-taking in Nuclear South Asia', CISAC Working Paper, Freeman Spogli Institute, Stanford University.

————. 2012. *Managing India's Nuclear Forces*. New Delhi: Routledge.

Krishna, G. 1984. 'India and the International Order: Retreat from Idealism', in H. Bull (ed.), *The Expansion of International Society*, pp. 269–87. New York: Oxford University Press.

Kux, D. 2001. *The United States and Pakistan, 1947–2000: Disenchanted Allies*. Washington, DC: Woodrow Wilson Center Press.

Ladwig, W.C., III. 2008. 'A Cold Start for Hot Wars? The Indian Army's New Limited War Doctrine', *International Security*, 32(3): 158–90.

Larson, T. 1969. *Disarmament and Soviet Policy, 1964–68*. New Jersey: Prentice Hall.

Levy, A. and C. Scott-Clark. 2007. *Deception: Pakistan, United States and the Global Nuclear Weapons Conspiracy*. New Delhi: Penguin

Malhotra, I. 1989. *Indira Gandhi: A Personal and Political Biography*. New Delhi: Hay House.

Massey, I.P. (ed.) 1991. *Nehru's Constitutional Vision*. New Delhi: Deep and Deep.

Mehta, J.S. 2006. *Negotiating for India: Resolving Problems through Diplomacy (Seven Case Studies 1958–1978)*. New Delhi: Manohar.

———. 2008. *Rescuing the Future: Bequeathed Misperceptions in International Relations*. New Delhi: Manohar.

———. 2010. *The Tryst Betrayed: Reflections on Diplomacy and Development*. New Delhi: Penguin, Viking.

Menon, R. 2000. *A Nuclear Strategy for India*. New Delhi: Routledge.

Menon, S.S. 2016. *Choices: Inside the Making of Indian Foreign Policy*. New Delhi: Allen Lane.

Ministry of Defence (MoD). 1968. *Annual Report, 1967–68*. New Delhi: Government of India.

Ministry of External Affairs (MEA). 1997. *Annual Report 1996–97*. New Delhi: Government of India.

———. 2005. 'Indo-US Joint Statement', 18 July. Available at http://www.mea.gov.in/bilateral-documents.htm?dtl/6772/Joint+Statement+IndiaUS, accessed on 10 January 2018.

Mistry, D. 2003. *Containing Missile Proliferation: Strategic Technology, Security Regimes and International Cooperation in Arms Control*. Washington, DC: University of Washington Press.

Mohan, C.R. 2003. *Crossing the Rubicon: Shaping of India's New Foreign Policy*. New Delhi: Viking.

———. 2006. *Impossible Allies: Nuclear India, United States and the Global Order*. New Delhi: India Research Press.

Mukerjee, D. 1968. 'India's Defence Modernisation', *International Affairs*, 44(4): 666–76.

Nagal, B.S. 2014. 'Checks and Balances', *Force*, 12–16. Available at http://forceindia.net/guest-column/guest-column-b-s-nagal/checks-and-balances/.

———. 2015. 'Strategic Stability—Conundrum, Challenge and Dilemma: The Case of India, China and Pakistan', *CLAWS Journal*, 1–22. Available at http://www.claws.in/images/journals_doc/1190813178_BalrajNagal.pdf, accessed on 10 April 2018.

———. 2016. 'India and Ballistic Missile Defence: Furthering a Defensive Deterrent', Carnegie Endowment for International Peace. Available at https://carnegieendowment.org/2016/06/30/india-and-ballistic-missile-defense-furthering-defensive-deterrent-pub-63966, accessed on 10 January 2018.

Narang, V. 2013. 'Five Myths about India's Nuclear Posture', *The Washington Quarterly*, 36(3): 143–57.

———. 2016. 'Strategies of Nuclear Proliferation: How States Pursue the Bomb', *International Security*, 41(3): 110–50.

National Archives of India (NAI). 1964. 'India and the Chinese Bomb: Note by KR Narayanan', MEA, File No. HI/1012 (14)/64, Vol. II, 26 November.

———. 1965. 'L.K. Jha to Prime Minister' (Top Secret), Prime Minister's Secretariat, File No. 30(36)/65/PMS, 23 March.

———. 1968. 'Annual Report for the Year 1967: Embassy of India, Peking (SECRET)', Ministry of External Affairs, File No. HI/1011 (5)/68, 24 April.

———. 1970. 'Brief on Government's Stand on the Resolution by Shri Virbhadra Singh, M.P. for Discussion in the House on 17th April, 1970', Department of Atomic

Energy (Top Secret), Prime Minister Secretariat, File No. 56/69/70–Parl, 24 April.

———. 1976. 'Pakistan's Capability to Produce Nuclear Weapons: Cabinet Secretariat, Department of Cabinet Affairs (Intelligence Wing)', MEA, File No. PP (JS) 4(3)/74 (Secret), 24 February.

———. 1977a. 'Pakistan-Clandestine Purchase of Nuclear Equipment and Materials (Cabinet Secretariat, Research and Analysis Wing)', MEA, File No. WII/504/3/77, Vol. IV (Secret), 29 September.

———. 1977b. NAI, 'Bhandari Additional Secretary (Foreign New Delhi) from Sood (Indembassy Paris)', MEA, File No. WII/504/3/77, Vol. IV (Secret), 10 August.

———. 1977c. 'Letter from Indian High Commissioner in Ottawa to JS (Pak)', MEA, File No. WII/504/3/77, Vol. IV (Secret), 18 July.

———. 1981. 'Pakistan's Nuclear Programme', MEA, File No. F/103/10/81 (Secret), 19 August.

———. 1982. New Delhi, 'US Policy Towards the Indian Sub-Continent: Note prepared by S. Jaishankar', MEA, File No. WII 104/16/82 (Secret), 11 February.

Nayar, K. 1987. 'We Have the A-Bomb, Says Pakistan's Dr. Strangelove', *Observer*. London, 1 March 1987.

Nayyar, B.R. 1977. 'India and the United States: New Directions and their Context', *Economic and Political Weekly*, 12(45/46): 1905–14.

———. 2001. *India and the Major Powers after Pokhran II*. New Delhi: Har Anand.

Nehru, J. 1986. *The Discovery of India*. New Delhi: Oxford University Press.

Nehru Memorial Museum Library (NMML). 1967a. 'Nuclear Policy', Prime Minister's Secretariat (Top Secret), IIIrd Installment, Subject File No. 111, 3 May.

———. 1967b. 'Nuclear Security', Prime Minister's Secretariat (Top Secret), P.N. Haksar Papers, IIIrd Installment, Subject File No. 111, 2 May.

———. 1967c. 'Letter from PN Haksar to Prime Minister Indira Gandhi' (Secret), P.N. Haksar Papers, IIIrd Installment, File No. 114, 7 August.

———. 1968. 'Instructions to India's Representative to UN on Non-proliferation Treaty' (Top Secret), P.N. Haksar Papers, I & II Installment, Subject File No. 35, 20 April.

———. 1974a. 'Ministry of Finance (Department of Expenditures), Revised Defense Plan 1974–79' (Top Secret), P.N. Haksar Papers, IIIrd Installment, Subject File No. 296, 25 December.

———. 1974b. 'Foreign Minister from Ambassador', T.N. Kaul Papers (As Ambassador to the US), I and II Installment, Subject File No. 1 (Personal), 2 August.

———. 1975. 'Comments of the Ministry of Defense on the Note Received from the Ministry of Finance' (Top Secret), P.N. Haksar Papers, IIIrd Installment, Subject File No. 296, January.

Noorani, A.G. 1967. 'India's Quest for a Nuclear Guarantee', *Asian Survey*, 7(7): 490–502.

————. 1978. 'Foreign Policy of the Janata Government,' *Asian Affairs*, 5(4): 216–28.

Norman, D. (ed.). 1965. *Nehru: The First Sixty Years*. Vol. II. London: John Day.

Oakman, D. 2010. *Facing Asia: A History of the Colombo Plan*. Canberra: ANU Press.

Ogden, C. 2014. *Indian Nationalism and the Evolution of Contemporary Indian Security*. New Delhi: Oxford University Press.

Pant, H.V. 2007. 'India's Nuclear Doctrine and Command Structure: Implications for Civil-Military Relations in India', *Armed Forces & Society*, 33(2): 238–64.

————. 2011. *The US–India Nuclear Pact: Policy, Process and Great Power Politics*. New Delhi: Oxford University Press.

————. 2015. *Handbook of Indian Defence Policy*. London: Routledge.

Pardesi, M. 2014. 'Chinese Nuclear Forces and Their Significance to India', *The Nonproliferation Review*, 21(3–4): 337–54.

Pedeo, W. 1976. 'Canada's Nuclear Power Program', *Ambio*, 5(3): 127–8.

Perkovich, G. 1993. 'A Nuclear Third Way in South Asia', *Foreign Affairs*, 91(90): 85–104.

————. 1999. *India's Nuclear Bomb: The Impact on Global Nuclear Nonproliferation*. Los Angeles: University of California Press.

Phalke, J. 2013. *Atomic State: Big Science in Twentieth-Century India*. Ranikhet: Permanent Black.

Press Information Bureau. 1998. 'Suo Motu statement by Prime Minister Shri Atal Bihari Vajpayee in Parliament on 27 May 1998'. Available at http://www.acronym.org.uk/old/archive/spind.htm, accessed on 10 April 2018.

―――. 2000. 'Sou Motu Statement Made in the Parliament by the Minister of External Affairs on the NPT Review Conference', 9 May. Available at http://www.acronym.org.uk/old/archive/spind.htm, accessed on 10 January 2018.

Raghavan, S. 2010. *War and Peace in Modern India*. Ranikhet: Permanent Black.

―――. 2013. 'The Fifty Year Crisis: India and China after 1962', *Seminar*, Vol. 641, January 2013. Available at http://www.india-seminar.com/2013/641/641_srinath_raghavan.htm, accessed on 10 April 2018.

Raina, A. 1981. *Inside RAW: The story of India's Secret Service*. New Delhi: Vikas Publishing House.

Rajagopalan, R. 2013. 'From Defensive to Pragmatic Multilateralism and Back: India's Arms Control and Disarmament', in W.P.S. Sidhu, P.B. Mehta and B. Jones (eds), *Shaping the Emerging World: India and the Multilateral Order*, pp. 197–216. Washington, DC: Brookings Institution Press.

Rajan, M.S. 1975. 'India: A Case of Power without Force', *International Journal*, 30(2): 299–325.

Ram, N. 1982. 'India's Nuclear Policy: A Case Study of the Flaws and Futility of Non-proliferation', Paper prepared for the 34th Annual Meeting of the Association of Asian Studies, Chicago.

Ramana, M.V. 2012. *The Power of Promise: Examining Nuclear Energy in India.* New Delhi: Penguin.

Ramanna, R. 1978. 'Peaceful Nuclear Explosions (PNEs)', in T.T. Poulose (ed.), *Perspectives of India's Nuclear Policy.* New Delhi: Young Asia Publications.

———. 2004. 'Five Decades of Scientific Development: Memories of PN Haksar', in S. Banerjee (ed.), *Contributions in Remembrance: Homage to PN Haksar* (Haksar Memorial Volume II). Chandigarh: Centre for Research and Industrial Development.

Rasgotra. M.K. 1991. *Rajiv Gandhi's World View*, pp. 59–72. New Delhi: Vikas Publishing House.

Reddy, E.S. and A.K. Damodaran (eds). 1994. *Krishna Menon on Disarmament: Speeches at the United Nations.* New Delhi: Sanchar.

Riedel, B. 2009. 'American Diplomacy and the 1999 Kargil Summit at Blair House', in Lavoy, Peter R. (eds). *Asymmetrical Conflict in South Asia; The Causes and Consequences of Kargil War*, pp. 130–43. New Delhi: Cambridge University Press.

Riqiang, W. 2013. 'Certainty of Uncertainty: Nuclear Strategy with Chinese Characteristics', *Journal of Strategic Studies*, 36(4): 579–614.

Salam, A. 1966. 'The Isolation of Scientists in Developing Countries', *Minerva*, 4(4): 461–5.

Sarkar, J. 2015. 'The Making of a Non-aligned Nuclear Power: India's Proliferation Drift, 1964–8', *International History Review*, 37(5): 933–50.

Schrafstetter, S. 2002. 'Preventing the "Smiling Buddha": British Indian Nuclear Relations and the Commonwealth Nuclear Force, 1964–68', *Journal of Strategic Studies*, 25(3): 87–108.

Science. 1956. 'Nuclear Progress in India', *Science*, 124(3217).

Sharma, G. 2013. 'The Making of Three Indian Nuclear Disarmament Plans', unpublished PhD thesis, Jawaharlal Nehru University, New Delhi.

Sheshagiri, N. 1975. *The Bomb! Fallout of India's Nuclear Explosion*. New Delhi: Vikas Publishing House.

Shukla, A. 2013. 'India's Missile Stories', *The Business Standard*, 20 September.

Sidhu, W.P.S. and C. Smith. 2000. *Indian Defence and Security: Industry, Forces and Future Trends.* Surrey: Jane's Information Group.

Singh, B. (ed.). 1988. *Jawaharlal Nehru on Science and Society: A Collection of Writings and Speeches*. New Delhi: Nehru Memorial and Museum Library.

Singh, J. 1998. 'Why Nuclear Weapons', in J. Singh (ed.), *Nuclear India*, pp. 9–25. New Delhi: Knowledge World.
———. 2006. *In the Service of Emergent India: A Call to Honour*. Bloomington: Indiana University Press.

Small, A. 2015. *The China–Pakistan Axis: Asia's New Geopolitics*. New Delhi: Random House India.

Sood, V.K. and P. Shawney. 2003. *Operation Parakram: The War Unfinished*. New Delhi: Vision Books.

Srinivasan, M.R. 2003. *From Fission to Fusion: The Story of India's Atomic Energy Program*. New Delhi: Viking.

Subrahmanyam, K. 1998a. 'India's Nuclear Policy: 1964–1998', in J. Singh (ed.), *Nuclear India*, pp. 26–53. New Delhi: Knowledge World.

———. 1998b. 'Nuclear Tests: What Next?', *India International Centre Quarterly*, 25(2/3): 50–62.

Subramanian, R.R. 1975. 'Nuclear India and the NPT: Prospects for Future?', *Instant Research on Peace and Violence*, 5(1): 35–9.

Sundaram, C.V., L.V. Krishan, and T.S. Iyenger. 1998. *Atomic Energy in India: 50 Years*. New Delhi: Department of Atomic Energy, Government of India.

Sullivan III, M.P. 1970. 'Indian Attitudes on International Atomic Energy Controls', *Pacific Affairs*, 43(3): 353–69.

Talbott, S. 2004. *Engaging India: Diplomacy, Democracy, and the Bomb*. Washington, DC: Brookings Institution Press.

Tasleem, S. 2016. 'Pakistan's Nuclear Use Doctrine', Carnegie Endowment for International Peace. Available at http://carnegieendowment.org/2016/06/30/pakistan-s-nuclear-use-doctrine-pub-63913, accessed on 10 January 2018.

Tellis, A. 2001. *India's Emerging Nuclear Posture: Between Recessed Deterrence and Ready Arsenal*. Santa Monica, CA: RAND Corporation.

Thakur, R. 1993. 'The Impact of Soviet Collapse on Military Relations with India', *Europe-Asia Studies*, 45(5): 831–50.

———. 1996. 'India and the United States: The Triumph of Hope Over Experience', *Asian Survey*, 36(6): 574–91.

The National Herald. 1964. 'Bhabha: India Can Make a Bomb in 18 Months', 5 October.

Thomas, R.G.C. 1980. 'Indian Defence Policy: Continuity and Change under Janata Government', *Pacific Affairs*, 53(2): 223–44.

Trivedi, V.C. 1975. 'India's Approach towards Nuclear Energy and Non-proliferation of Nuclear Weapons', in A.W. Marks (ed.), *NPT Paradoxes and Problems*, pp. 42–6. Washington, DC: Carnegie Endowment for International Peace.

United States Arms Control and Disarmament Agency. 1968. 'World Military Expenditures 1966-67', Research Report 68-52, p. 16. Washington, DC: Government of the USA.

Venkataraman, R. 1998. 'Nuclear Explosion and its Aftermath', *USI Journal*, CXXVIII(533): 303–9.

Walker, W. and M. Lonnroth. 1988. *Nuclear Power Struggles: Industrial Competition and Proliferation Control*. London: George Allen and Unwin.

Wit, D. and A. Cubock. 1958. 'Atomic-Power Development in India: Prospects and US Role', *Social Research*, 25(3): 285–302.

# Index

# About the Authors

**Harsh V. Pant** is distinguished fellow and head of strategic studies at Observer Research Foundation, New Delhi, India. He holds a joint appointment as professor of International Relations in Defence Studies Department and the India Institute at King's College London, UK. He is also a non-resident fellow with the Wadhwani Chair in US–India Policy Studies at the Center for Strategic and International Studies, Washington, DC. Pant is also a columnist for the *Diplomat* and writes regularly for various media outlets including the *Japan Times*, the *Wall Street Journal*, the *National (UAE)*, and the *Indian Express*.

**Yogesh Joshi** is a Stanton Nuclear Security postdoctoral fellow at Center for International Security and Cooperation, Stanford University, California, USA.

He has a PhD in international politics from Jawaharlal Nehru University specializing in Indian foreign and security policy. He has held fellowships at George Washington University, Washington, DC, USA, King's College London, UK, and Carnegie Endowment for International Peace, Washington, DC, USA. He has co-authored two books: *The US 'Pivot' and Indian Foreign Policy: Asia's Emerging Balance of Power* (2015) and *India and Nuclear Asia: Forces, Doctrines and Dangers* (2018).